BEEKEEPING
For
BEGINNERS

........Build Thriving Hives, Harvest Honey, and Grow Your Own Bee Empire

MOBAROK HOSSAIN

Copyright © 2024 by **Mobarok Hossain**

All rights reserved. No part of this book may be reproduced, distributed, or transmitted in any form or by any means, including photocopying, recording, or other electronic or mechanical methods, without the prior written permission of the author, except in the case of brief quotations embodied in critical reviews and certain other noncommercial uses permitted by copyright law. For permission requests, write to the author at the address below.

CONTENTS

INTRODUCTION 9

CHAPTER 1 17
    HONEYBEE ESSENTIALS 17
    The life cycle of a honey bee 17
    The Roles With in a Hive: Queen, Workers, and Drones 21
    Understanding Bee Species 22
    The Role of Bees in Pollination 25
    More than Just Honey: The Surprising Impact of Bees 27

CHAPTER 2 29
    GETTING STARTED IN BEEKEEPING 29
    Legal and ethical considerations 32
    Starting Your First Hive: Key Decisions 34
    Finding a mentor or beekeeping community 36

CHAPTER 3 39
    ESSENTIAL EQUIPMENT AND SAFETY 39
    Honey Extractors 42
    Protecting yourself: suits, gloves, and more 43
    Hive Inspection Guidelines 44
    How to Manage Your Hive Safely and Efficiently 46
    Best Practices for Bee and Keeper Safety 47

CHAPTER 4 49
    CHOOSING THE RIGHT HIVE 49

Understanding hive types 50
Setting Up Your Hive 55
Hidden Components and How They Work Together 58

CHAPTER 5 61
OBTAINING AND INTRODUCING YOUR BEES 61
Buying vs. Capturing Bees: Pros and Cons 62
How to Address Common Challenges 68

CHAPTER 6 71
SEASONAL HIVE MANAGEMENT 71
Spring: Preparing for Growth 72
Swarm Prevention 73
Summer: Managing expansion and honey production 74
Steps for Harvesting Honey: 75
Maintaining Hive Health 76
Fall: Preparing for Winter 77
Types of Fall Feeding: 78
Winter: Overwintering Your Colony 79
Keeping track of hive activity 80
Important Things to Note During the winter 81

CHAPTER 7 83
HARVESTING HONEY AND OTHER HIVE PRODUCTS 83

Introduction: The Sweet Rewards of Beekeeping 83

How to Harvest Honey the Right Way 84

Extraction techniques for every beekeeper 84

Tips from Expert Beekeepers: Maximizing Your Harvest 90

CHAPTER 8 93

BEE HEALTH AND HIVE MANAGEMENT 93

Managing hive pests and preventing infestations 98

Maintaining healthy bees through nutrition and care 101

Water and hydration 102

CHAPTER 9 105

THE ROLE OF BEES IN POLLINATION AND BIODIVERSITY 105

Why bees are critical pollinators 106

Using Your Bees for Pollination Services 108

Creating a pollinator-friendly garden 110

Beekeeping and conserving biodiversity are synonyms 111

Illustrative Case Studies 112

CHAPTER 10 115

TROUBLESHOOTING COMMON BEEKEEPING ISSUES 115

Swarming: How to Prevent and Manage It 116

- Identifying the causes of aggression 118
- Dealing with and Soothing Bees in a Non-Threatening Way 119
- Addressing and solving practical problems 120
- Pests and environmental stressors 121

CHAPTER 11 123
- BEYOND HONEY: EXPANDING YOUR BEEKEEPING HORIZONS 123
- Splitting hives and queen rearing 124
- The Art of Raising Queen Bees 126
- Turning beekeeping into a business 128
- Honey marketing and product diversification .. 129
- Product diversification in a honey business includes the following: 130

CONCLUSION 133
- THE JOURNEY OF BEEKEEPING: FROM HOBBY TO PASSION 133
- Coming to embrace the Beekeeping Journey ... 136
- Concrete steps toward environmental responsibility 137
- Look Ahead 138

GLOSSARY 140
- KEY TERMS USED IN THIS BOOK 140

INTRODUCTION

Historically, beekeeping was perhaps centered on providing additional food sources or enriching a family's resources. It has now developed into a hobby for some, a business for others, and a lifetime obsession for many. People bridge the gap between themselves and nature in quite a special way. Today's title, Beekeeping for Beginners: Build Thriving Hives, Harvest Honey, and Grow Your Own Bee Empire, is all inclusive where it ensures that you do not miss on any stage of the beekeeping adventure. Aside from producing honey for yourself or for sale, helping Mother Earth, and knowing about bees, then you came to the right place. The first part of this chapter addresses the question of why beekeeping is of interest to people and what kind of pleasure Russian honey hunters extract from this pastime.

1. The fascination of beekeeping

While many may think of it as a different type of gardening or farming, beekeeping has the hems of a different hobby. After all, stinging insects are not so many people's favorites. The quick realization by those who venture into beekeeping is that the bees are not the only things to be kept. It is repurposed from working with the landscape to feeling the weather cycle and observing a colony that functions as if someone has perfected coordinated motion.

It is amazing to think how, with enough effort, even simple contents such as flower nectar can be transformed into honey, beeswax, and propolis, which in turn are of very high relevance for man. There's beauty and fascinating experience that makes one forget the zones within which we occur, for instance, while hundreds of bees construct and perform their respective duties, or while several people from the same community collaborate and build something bigger than themselves. Every time it is one more marvel, from noticing how the bees construct their hives and arrange the ink industry within them to the sight of their dances and smell-intense communication.

There are also components of management with beekeeping. The productive action is not one-sided, as one is raising bees here, but also an environment where these bees can be successful. It is fulfilling to raise a colony of bees, face the adversities and challenges of beekeeping, and in the end witness a grown and thriving colony that you have nurtured. It is amazing how the effects of bees on a person can be rather calming, and even for a season's end, there is that sweet benefit of honey that elevates the pleasures.

All these and other reasons are the determinants that pull some beekeepers today. With bee populations lacking healthy ecosystems to survive, they become engaged in beekeeping. This personal contribution to sustainability is both meaningful and motivating for new and experienced beekeepers alike.

2. Why do bees matter to the world?

Bees are blatantly neglected as honey producers only. However, their primary contributing role is as pollinators. Bees account for the pollination of 75% of the world's fruits, vegetables, nuts, and seeds, which are food crops. If these little humming workers were to cease existing, a lot of the food we enjoy today would either be very limited or nonexistent at all. The science and practice of farming also depend on bees, which help grow apples, almonds, blueberries, and coffee.

The importance of bees in promoting ecological diversity is beyond description. Pollination of plants leads to sustaining biological diversity in various ecosystems through the action of bees. When bees pollinate wild flowers, they help maintain wild flower beds, which act as homes for many creatures, including insects, birds, mammals, and others. Without bees, some food chains or web structures will be removed or wiped out from the boundary, and so many wild creatures in this case will be on extinction, and the end results for humans will be hazardous to say the least.

But still, bees seem to be under siege now more than ever. Depletion of their habitat, adverse climatic conditions, use of pesticides, and epidemic conditions such as the bee colony collapse disorder have affected a good number of bees. This has raised alarm levels internationally, with beekeepers, farmers, scientists, and environmental care combining efforts to come up with solutions. By establishing your own hive, you are becoming part of a movement that is key to transforming our agricultural practices and the environment.

3. What you'll learn from this book

This guidebook will give every beginner in beekeeping a sound grasp of bees and beekeeping skills. The trip you are going to take is a tutorial and a practical one, and this book will simplify things for you by guiding you through every process starting from the first hive setting to honey production and bee health control.

A summary of the topics discussed

- **The Life of a Honey Bee:** This chapter takes you through the life stages of honey bees (bee forms—from an egg to an adult bee) and the caste system of queens, workers, and drones as all divided how they work together.

- **The Different Types of Bees and Hives:** You will differentiate the various types of honey bee species and honey bee behaviors together with the selection of the appropriate types of bees and hives. We'll also look at different approximation hives, such as the miscellaneous Langstroth, Warre, and top-bar hives.

- **Pollination and Its Importance:** You will learn that there is a process in which bees contribute, and that is in patient's gardens and in pollination.

- **Hive Maintenance and Seasonal Care:** Beekeeping is not a hobby activity that should be done once in a long time. You will be taken through managing your hives during the different seasons of the year while keeping your bees productive and in good condition.

- **Using the Hive:** Everything from harvesting honey and beeswax to how you can use the propolis your bees produce.

- **Preventing and Managing:** Know the early signs and symptoms of common bee health problems, and find out what to do about pests in your beehive.

4. The Benefits of Beekeeping: A Personal and Global Perspective

From a personal perspective, one can find a lot of inner peace when beekeeping. There is this wonderful feeling in taking out the honey that has been produced by the bees that you have nurtured. It's an opportunity to take a break, enjoy the environment, and learn what's going on in the world even better. The majority of the beekeepers describe this activity as peaceful and relaxing, away from the pressure of normal life.

On the economics aspect, the advantages are even more staggering. In the course of keeping a beehive, one supports the environment as well as the increase and development of habitat. They will assist in the pollination of fruits, plants, shrubs, and crops, enhancing food security and diversity. At a time when there's a threat that bee populations will eventually become extinct, beekeeping is key to helping restore the health of both pollinators and the ecosystem.

At the same time, it is worth mentioning that beekeeping practices may have income opportunities as well. There is great demand for honey and even more other bee products such as tops, propolis, and jelly. The majority of beekeepers make a hobby into part-time businesses and so sell honey at the farmers market or perhaps candles made from beeswax, or offer pollination to farms.

Bee-keeping A fun way to meet people in your community, help the environment, and save some money The film shows us the bigger picture of bees in your life, but it extends into a story about bees in life so much beyond our own, and it offers us specific points from where we can be of help.

CHAPTER 1

HONEYBEE ESSENTIALS

While bees in general are such fascinating animals that are so useful in physical and economic domains on the entire planet, their immediate relatives within the internal environment of the honeybee colony can be quite off (in this case, the society of other bees). Exploring these aspects and many more will facilitate one's understanding of beekeeping. It is equally important to know that there are also many other kinds of bees aside from honey bees, each type possessing a set of attributes that makes it suitable for a given application. Lastly, we will also discuss the importance of bees in pollination and the many other benefits of beekeeping apart from harvesting honey.

The life cycle of a honey bee

Honeybees follow a structured life cycle where every stage is crucial for the running of a hive. The cycle includes the following stages: embryo stage, larval stage, worker stage, and other adult bees.

After 21 days of development, worker bee tenders can go from the egg to the adult stage, after which the duration may shorten depending on the type of bee.

After the egg is mature and she goes to a creche, she joins a laiding course for one year.

- **Egg Stage:** The queen bee lays eggs, say, between one thousand five hundred (1500) and two thousand (2000) when she is at the top of her game. Each fertilized egg is put inside a wax compartment or cell in the center of the hive. The egg is then developed into a larva on day three.

- **Larva Stage:** During this stage of growth, the larvae are fed only royal jelly or a mixture of royal jelly and insect honey bee bread, which consists of pollen and nectar, which has been prepared by the worker bees. The hurt will also tend to these larvae and feed them until they are grown up. There is rapid larval growth during this period, which is why they also go through several molts.

- **Pupa Stage:** A period of six days passes; a worker nurse occupied with the larvae begins the process of pupation by drawing the lid or the cap of the cell. Within the cell that has been covered, the larvae spin silk and change into a butterfly with wings, legs, and eyes, among other adult characters. This stage lasts about twelve days.

- **Adult Bee:** Once the pioneers of the metamorphosis are over, the new, already developed adult bee digs out through the inner wall of the cell and flies out as an imago. The bee could either be a worker, a drone, or a queen, depending on her function in the hive.

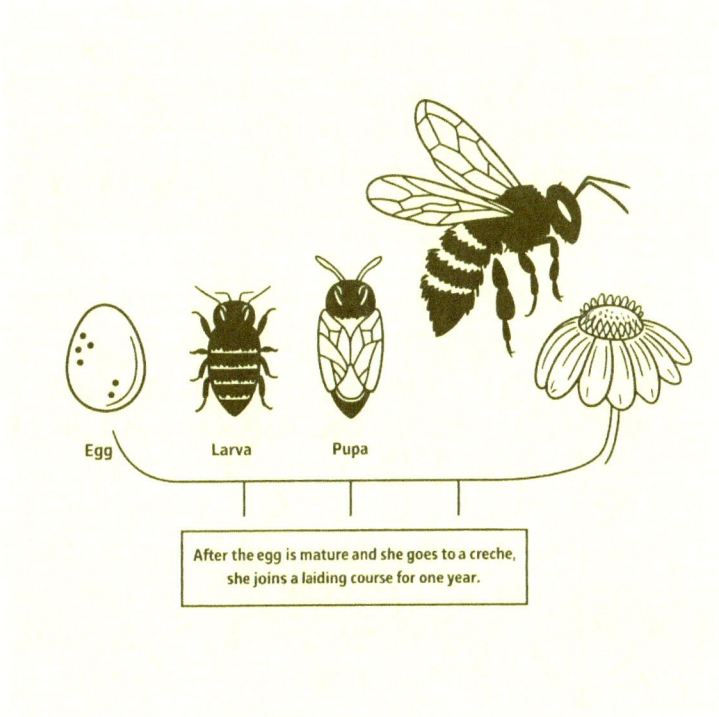

Egg   Larva   Pupa

After the egg is mature and she goes to a creche, she joins a laiding course for one year.

How bees communicate and navigate

One of the most fascinating aspects of honeybees is the way they coordinate and perform certain complex activities with a sense of accuracy. While communicating, players wear and deploy two types of communication: the primary, pheromones, and the second, the "waggle dance."

- **Pheromones:** Not only do bees use pheromones for those activities, such as mating with a partner, but they also use them to relay instructions. In this case, the pheromone signals the workers that the queen is fine and continues to make eggs. Bees in the foraging process do also emit particular alarm pheromones that they use to warn the members of the hive of any perceived dangers.

- **The Waggle Dance:** When there is food in the vicinity, the honeybees move in a diminishing circle, which at last and invariably leads to the vertical position of the bee's body. This is known as the waggle dance, which serves to communicate the position of the food or nectar for the honeybees. A particular bee is able to tell how far away a flower is by performing the 'waggle' for how long compared to other movements of her body. Placement of the bee in relation to the sun indicates the direction to take, while pressing for the waggle indicates the extent of the distance to be covered. It is one of the best ways motivational systems operate to ensure that foraging clears some of the more scrumptious food sub-accounts.

The Roles With in a Hive: Queen, Workers, and Drones

A well-functioning hive operates like a highly organized society, with each type of bee playing a specific role.

- **Queen Bee:** The queen bee is the only sexually mature female in the colony. He is responsible for approximately 99.9% of the colony's reproduction by laying her eggs inside the hive. A queen is capable of laying ovum hundreds and even thousands of times within one day, and she is the one responsible for emitting the queen signal pheromone. She also makes a decision on whether to fertilize the eggs, whose potential is used to develop into nurses, or not, and such eggs become non drones.

- **Worker Bees:** Worker bees are non-reproductive females and make up the majority of the hive structure. Their roles are not static but change with their age, initially as nurse bees that nest and cluster around the larvae, then as foragers for nectar and pollen, clean the hive, make honey, and defend the hive. They form the very center of the colony, if not the glue, undertaking almost every function required for the wellness of the colony.

- **Drones:** drones are the male bees. Hence their function is limited to copulating with a queen bee. They do not feed or protect the hive. Drones die after mating, and those that do not mate generally face expulsion from the hive when winter comes to preserve the workloads of bees.

Understanding Bee Species

Introducing bees for the first time requires that the beginner appreciate the types of bees and their respective characters. The bees that you select to use for your apiary will possibly determine the health of the colony, the quantity of honey that you harvest, and the ease of the bees' management.

The Western honeybee (Apis mellifera)

By far the most popular bee farming in the world is known to keep the traditional species Apis mellifera because it gives honey, pollinates, and has a calm temper. They are so well adapted to diverse climatic regions, making this a boon to beekeepers in different parts of the world.

Western honeybees are known for their:

- **Productivity:** They produce honey at an exceptional rate, which makes them the most preferred species in honey production.

- **Temperament:** Western honeybees are less aggressive compared to other kinds of species, and so beginners find it easier to handle them.

- **Adaptability:** These bees do so well in different weather and climatic regions, such as temperate and warmer weather; hence, they are ideal for beekeepers in various places.

Africanized Bees: Separating Fact from Fiction

As the name suggests, Africanized bees (also known as killer bees) are the cross of the western honey bee and the African bee. They were inadvertently brought to America in the biological invasion in the latter half of the 20th century. Nonetheless, it is very vital to note that although such a bee borrows the famous 'killer bee' thumbs up, there is more fiction than fact.

- **Aggressiveness:** With respect to sexual ratio, it is an amplified realistic picture that Africanized bees tend to be more defensive of the hive compared to western honey bees when the latter bees are interrupted in the

comb. Only exceptions are the induced methods, where bees become aggressive after exposure to novel types of meteorological conditions.
- **Suitability for beekeeping:** As for Africanized bees, these species should not be the first priority for newcomers since they are less tolerant of bee hives in general than the rest. They are fine for experienced beekeepers, but only unless all safety precautions and extra measures are directed towards the use of the bees.

Choosing the Best Species for Your Apiary

The Western honeybee (Apis mellifera) is the most popular choice for the novice beekeepers. They are relatively simple to handle and are able to adapt quite well to several areas. However, if you are in the region where Africanized bees are found, it's important to know the local laws and how to manage your hives to reduce the chances of hybridization.

The Role of Bees in Pollination

Pollination is another area where the bees remain quite tactically important. Pollen is transferred from one flower to another as bees gather nectar and pollen from the flowers, helping in fertilization. This process of fertilization leads to the maturity of plants, fruits, seeds, and vegetables, which is important for food and the biosphere.

How Bees Help Feed the World

Global hunger is a problem that has many dimensions, and Part of the dimension of this work is how quite bees feed the world, doing the toughest job like pollination. Approximately 75% of the crops produced by humanity make any food without pollinators, particularly bees. Apples, almonds, blueberries, and cucumbers are common examples of crops that depend on bee pollination. Most of our delicious foods would not be that abundant without bees, which makes food security a real problem.

Exemplifying their ecological importance, bees not only enhance agricultural productivity but also contribute to the ecological sustenance of the wild flora. These plants provide food and habitat to various species of animals, aiding in the conservation of biological diversity.

The Science of Pollination and Biodiversity

When bees pollinate plants, they contribute to the genetic diversity of plant species, which is critical for the health of ecosystems. Pollination increases the resilience of plants, allowing them to adapt to environmental changes. This biodiversity supports healthy soils, water systems, and habitats for countless species, all of which are essential to human well-being.

More than Just Honey: The Surprising Impact of Bees

Despite the fact that the production of honey is one of the most known agriculturally beneficial roles of the aforementioned species, this role far exceeds the sweet banana. There are bee products other than honey, such as bee wax, which is used in several products, including beads and cosmetics. Another bee product, propolis nanoparticle, collected from tree resin and used in exotic medicine, is claimed to be beneficial.

Apart from the economic benefits realized from bees' products, there are other social benefits derived from their activities. The plants that result from this particular activity, offering food and shelter to other animals, enhance the whole ecology of other animals, inclusive. It is, however, worth nothing that keeping he bees helps one not only support the environment but also contributes to the conservation of biodiversity and the availability of food.

CHAPTER 2

GETTING STARTED IN BEEKEEPING

Though exciting and fulfilling, beekeeping is a venture that requires a thorough grasp of certain elements before starting off. This time, you will learn all the basic information on beekeeping. It encompasses basically all aspects, from understanding the basic knowledge that is required to battling some of the most common beginner's barriers. For this reason, we will offer you all the necessary steps to getting started. In addition, we will address the many legal and moral issues that you have to consider, how to build positive relationships with your neighbors, and major choices that you will face as you prepare to establish your intercepting hive. Finally, we will emphasize the importance of finding a beekeeping community or experienced beekeeper to mentor you.

No matter what the reason to obtain honey, to help nature, or just for pleasure from keeping these wonderful insects—this chapter will provide the necessary background to be a professional in beekeeping. So let's go ahead and jump right in!

The learning curve: what every beginner needs to know

It should be pointed out that although beekeeping is a very rewarding hobby, there are a number of things that every beginner must come to terms with too. The beauty of watching bees and seeing them work and do all the luscious honey is both tantalizing and captivating, and yet working and managing a hive is not all about what is seen.

Here are the key things to keep in mind as you embark on your beekeeping journey:

- **Essential Beekeeping Knowledge**

    Before it is time to get your first set of bees, understanding the biology and management of bees as a beginner is crucial. It will help to learn how a honeybee colony works and the duties of the queen, worker bees, and drones, as well as how all the different life cycles of the bees are. One will also need to know the different parts that...how honey is produced by bees and how healthy colonies can be maintained.

You will find that terms like 'brood,' 'honey supers, 'brood chamber,' and' swarming' will be alien no more. These new terms have to be studied and understood early to boost your confidence when working on your bees. Another important capability is knowing how to examine your hive. Standards and techniques need to be put in place to allow for examination of the bee colony over time. Simple procedures will help to cure brokers and check for other diseases of pests. You will check the most visible sign of health, whether your queen is there or not

- **Navigating the Challenges**

While there are positive aspects of beekeeping, it does have its fair share of drawbacks. One of the greatest challenges is assessing the natural habits of bees. For instance, there is an instance where bees decide to make swarms due to overpopulation. Or during hot days, the bees may become hot-tempered. This is a skill learned with experience: noting the mood of the hive and what they are requiring.

Another problem that has been noted is the management of pests and diseases. New bee keepers come across pests such as varroa mites, small hive beetles, and foulbrood, which are some of the common pests to the bees. Be ready to learn about treatment and pest management methods that depend on the location and problem.

Refraining from this is another obstacle. There is a long-term responsibility and tenderness that comes with a beekeeping activity. One has to master the customs of their hives, from feeding them to making all the appropriate literary decisions about honey production.

Legal and ethical considerations

- **Beekeeping laws and regulations**
  Researching the local laws is often beneficial before embarking on your beekeeping adventure. Certain jurisdictions have regulations concerning capitals that control the sites for hives, with respect to the number of hives allowed per house as well as what species of bees are permitted. Some states or even countries will ask you to get your hives registered and/or some procedures of inspection followed to keep the health of your bees and control diseases.

It is imperative that these precautions be observed, not only for the well-being of your bees but also for the health and wellbeing of your neighbors and the ecosystem. These laws uphold the dignity and long-lasting existence of beekeeping as a practice in the economy.

- **Neighborhood relations and public education**

    One of the most neglected of the beekeeping practices is encouraging positive neighbor relations. While you may love bees, most people are afraid to live near beehives. Speak to them and have them understand exactly how bees benefit their environment, pollinate flowers, and the usefulness of such a tiny insect.

    Hive placement should also prevent lessees from disturbing neighbors when they use their homes. A hive, when properly placed, is in a quiet place where people don't walk around most of the time, making the worry that bees disturb a lot of people irrelevant.

You can also turn people who might be worried into those who will appreciate bees by giving such people some fresh honey from the bees.

Starting Your First Hive: Key Decisions

- **Choosing Your Hive Location**

    One of the most important steps is choosing where to place their hive. Bees prefer warm and sunny areas, which are also protected from strong winds. You should not situate your hive in very shaded locations, as bees require some heat, and the rightly situated hive will promote maximum bee activity all through the seasons.

    Make sure the bees are well-hydrated, but don't put anything near your house or your neighbor's yard that could bother people. Something as simple as a small birdbath or a low-water container can work perfectly for this purpose.

    Moreover, think about accessibility for yourself. Join me on a trip to your hive. You will be undertaking many visits to your hive. Therefore, carefully consider the location of the hive so that inspections, honey harvesting, and hive maintenance will not be cumbersome.

- **Buying vs. Catching Your First Bees**

    Now that your hive is complete, it is time to put in your bees. The first colony may be acquired in two principal ways: acquiring bees or rather catching them.

    For novices, it is easier to buy whole colonies of bees than any other method. Bees can be obtained in the form of a "nuc" (nucleus colony) or as colis of bees from a reliable supplier. It contains a queen, worker bees, and a few comb frames that have brood on them, which makes it ideal for novices since it's almost ready to use. A bee package on its part contains mother brood and disorganized adult bees and calls for some initial arrangement, which is still helpful for beginners.

    Catching a wild swarm is another option, but it's more challenging and requires experience. Swarming happens when bees naturally leave their hive to establish a new colony. Catching a swarm is exciting, but it's risky for beginners because you need to handle the bees carefully and ensure they're healthy.

    Catching a wild swarm is another option, which, however, is more difficult and calls for some experience. Swarming occurs when the bees of a colony decide to leave that colony in search of a place to settle, and a new home is developed. There is a lot of thrill in attempting to catch a swarm, although it is not recommended for beginners as it presents some dangers in that people should be careful to capture the bees and ensure they're healthy.

Finding a mentor or beekeeping community

The Power of Joining a Beekeeping Club

If the thought of keeping honey bees sounds super daunting, well, the good news is that you are not left to do it by yourself. The best method of knowing is to become a member of a beekeeping society or community in the area. Who can offer advice, answer questions, and provide support as you get started.

With a beekeeping club, one will be apprised of the factors that have the potential of affecting one's bees, e.g., disease or war. A large number of societies also conduct amazing applied activities whereby members get an opportunity to rehearse aspects such as looking after honey bees and getting honey before actually doing it on their personal colonies.

Also, you can search for mentors because they are quite important in enhancing your beekeeping journey. A mentor can provide you customized solutions to your challenges and recommend practical ways to avoid some of the errors. For example, they can take you through your very first hive inspection, teach you what a thriving (or struggling) colony looks like, and dispense useful tips acquired over the years.

Working with a mentor provides invaluable insights that books or videos can't always convey, and having someone to guide you builds your confidence in managing your bees.

CHAPTER 3

ESSENTIAL EQUIPMENT AND SAFETY

Taking care of the honeybees is one of the most enjoyable tasks, although you need to be equipped well and know the safety measures. In this chapter, the author will discuss specifics of the most important equipment you will need, how to inspect your hives, and how to keep the two best things about beekeeping you and your honeybees safe.

Beekeeping Tools You Can't Live Without

At the beginning of every beekeeping beginner finds themselves daunted by the flood of all those alternate instruments. Some of the instruments we all know the words to can't be spared by any beekeeper, no matter whether he/she is a greenhorn or an experienced professional. Below, we outline the essential tools you will need.

Hives

The hive is where your bees reside, and it is probably the most important element of your beehive machinery. Below are common examples:

- **Langstroth Hive:** They are the most commonly used hive styles. They consist of vertically stacked boxed units that facilitate easy bee management. They can be opened up during the removal of frames to allow access for honey extraction and to control the bees.

- **Warre Hive:** This is a low-handed bee house that copies more of the natural mode of nesting in bees. It is less complicated than the Langstroth. Good for the people whose desire is more of a cool approach to beekeeping.

- **Top Bar Hive:** This hive has a more horizontal construction and helps bee colonies to naturally build their combs. It is suitable for even children and is less complicated but may not produce as much honey as more advanced properties.

Smokers

A smoker is one of the most important pieces of equipment when inspecting the bees, as it serves to pacify them. When honeybees smell smoke, they prepare themselves to leave the hive; hence, they become less furious. This is how you use a smoker:

- Start by stuffing the smoker with heat-sensitive materials such as pine needles, burlap and thick cardboard.

- Ignite the above materials and the moment they begin to create smoke, pumping the bellows will ensure that the smoke is continuously released.

- Smoke should then be applied lightly at the entrance of the hive and the hive interior where the inspection is to be carried out.

Hive Tools

These are professional items that assist you in the management of your hives. Important tools include:

- **Hive Tool:** This flat metal tool is particularly designed to separate the frames and boxes of the hives, remove propolis, and do a number of other activities around the hive.

- **Bee Brush:** This is a brush that contains soft hair and is applied to bees to help remove them from frames or devices without harming them.

- **Frame Gripper:** This is some sort of tool that enables the user to remove frames from the hive without too much force, which may result in the squashing of bees within the frames.

Honey Extractors

A honey extractor will be necessary if you want to harvest honey. It lets you remove honey from frames without damaging the comb. The two most important categories are:

- **Manual Extractors:** These are used from the very first stages and foster small activities.

- **Electric Extractors:** These are used with the power and process honey in larger quantities comparatively faster and hence used in large beekeeping operations.

Protecting yourself: suits, gloves, and more

When it comes to bees, safety should always be given a priority. Below are the major protective gear pieces that you need to have:

- **Bee Suit:** A full bee suit works to protect every part of the body from any sting. Get a suit that allows you to move fluidly and is made of light material.

- **Bee Veil**: This is a mesh screen that protects your face and neck from stings while allowing you to see. Some veils come attached to the bee suit for extra safety.

- **Gloves:** Long gloves that are covered with leather or any tough fabric remain hard but have some movement. Gloves should be tightly fitted to prevent stings when on the hands.

- **Boots:** It is also common sense to wear long boots with an airtight top to prevent bees from crawling in.
- **First Aid Kit:** Always have a first aid box. You can also carry anti-allergy tablets, cold therapy packs, and hydrocortisone cream for the treatment of bee stings.

Hive Inspection Guidelines

Conducting thorough health checks should be a regular practice in order to ensure that bee colonies are in a healthy state. Having an understanding of what to check in every season will help you as a hive manager.

Things to Be Consistent and Check in Every Season

Spring

- **Queen Activity:** Look out for any queen egg deposition. Check for the presence of brood and eggs in the cells.

- **Population Growth:** Evaluate the number of bees in the colony in relation to space available for growth.

- **Food Supply:** Make sure there are enough stores for the bees as they come out of dormancy for winter.

Summer

- **Honey Production:** Inspect the run honey and see the process of filling with honey by the bees. If hawneys are completely empty, it may be due to other reasons.

- **Swarm Prevention:** Attempt to prevent swarming and therefore decrease bee populations by looking for signs of overcrowding or queen cells.

- **Pests and Diseases:** Carry out inspections to identify the presence of parasites such as the Varroa mite and any diseases, including American foul brood.

Fall

- **Honey Harvesting:** If you have not started harvesting honey, then it is the time to do so.

- Leave behind sufficient honey to enable bees to survive through the winter.

- **Winter Preparation:** Check for enough food and buy sugar if there is a likelihood of food deficiency.

Winter

- **Hive Stability:** Ensure the hive is fastened and does not fall into strong winds.

- **Moisture Management:** Check that moisture doesn't accumulate inside the hive, which can harm the bees.

How to Manage Your Hive Safely and Efficiently

- **Schedule your inspection:** It is prudent to set a time to conduct a review of the same hive. Prefer every 7-10 days if it is the peak season.

- **Keep it calm:** Stepping into the vicinity of your hives is a slow and methodical approach. Do not do anything hurried, as this will alarm the bees.

- **Work During Optimal Times:** Conduct hive examination during the warmest part of the day when bees are most active.

- **Minimize Open Time:** The deeper the hive is opened, the greater the chance of causing trauma to the bees. Hence, keep the hive open for the least time possible.
- **Limit Your Noise:** When talking around the hive, use a low tone and refrain from producing any noise that might provoke bees.

Best Practices for Bee and Keeper Safety

For a successful apiculture process and safety, there are some practices that should be followed for both bees and beekeepers:

- **Know Your Bees:** Spend some quality time with your bees, and you will come to comprehend their behavior. This way, you will be able to know what to expect from the bees and avoid being stung.

- **Use Smoke Wisely:** Fire may be effective in quietening bees; however, it is good to be economical when using it. Excessive use of smoke may wear them out and make them aggressive.

- **Stay Calm:** All the same, should one find himself within a swarm of bees that is agitated, it is good not to panic. People who are not aware of this may overly run or make quick motions, which they are not supposed to do. If she has been flying away from you and finds herself being pursued, then you simply have to walk away.
- **Have a Plan for Allergies:** Make sure that any possible allergies to bee stings, both for yourself and those around you, have been planned for. To take precautionary measures, if you or anyone else is in danger, an epinephrine auto-injector should be kept close.

- **Educate Your Community:** It is also vital to educate peers, friends, and other members of the community to win hearts and understand beekeeping. Invite them home to see your colonies and make them feel at ease with the offspring.

- **Practice good hygiene:** Before leaving your apiary or after coming from one, wash your hands immediately. This prevents any infections that you may have caught from the bees infecting you back.

- **Regularly Inspect Equipment:** Go through the storage compartments of your bee equipment to make sure that no devices or essential tools are missing. Ensuring that tools are regularly maintained helps to keep them in good working order and safe.

CHAPTER 4

CHOOSING THE RIGHT HIVE

Selecting the perfect hive is one of the most crucial choices that needs to be made by a beekeeper. Which is why the specifics, like the overall health of the bees, how much honey is produced, and general organization tend to vary depending on what kind of hive was chosen. To enhance the success of beekeeping, there are several hive types largely used, and understanding both the merits and demerits of each is important.

This chapter will present you with the various hives available for beekeeping and particularly examine the three most famous ones, i.e., the Langstroth hive, the Warre hive, and the Top Bar hive. These three beehives each offer distinct characteristics, benefits, and challenges that suit particular styles of beekeeping. Also, we will go over the basics of the hive installation process, its correct location, placement, orientation, parts of the hive, and their effective interactions. Eventually, we shall explore some alternative current beehive designs, such as the self-draining hives, that are likely to change the face of beekeeping.

At the end of this chapter, you will have mastered the means by which you will be able to select a beehive that meets your requirements and know how to set it up in an efficient manner in preparation for the hives perfect use.

Understanding hive types

Langstroth Hive: The Beekeeper's Favorite

The Langstroth hive is the most widely used in the world, and for good reason. Designed by Reverend Lorenzo Langstroth in the mid-1800s, this hive incorporates the concept of "bee space," allowing bees to move freely between frames. Here's a breakdown of its benefits and drawbacks:

Why the Langstroth hive is the most popular worldwide is clear. It was invented in the mid-1800s by the Reverend Lorenzo Langstroth and incorporates the idea of "bee space," which makes it easier for the bees to traverse frames. Let's go through the advantages and disadvantages of it:

Benefits:
- **Modular Design:** The Langstroth hive consists of components that make it up into boxes (supers); such supers can be added when necessary, and when the hive is no

longer needed, most of the supers can be removed. This makes it handy to increase in the hive size whenever the size of the bee population increases.

- **Ease of Management:** The structure of the hive also permits easy examination. It thus does not disturb the whole colony, for it is easy for the Keeper to reach the frames with little regard to the colony.

- **High Honey Production:** The advantage of adding supers means that this is the best method in terms of honey harvesting, and hence most of those interested in honey production prefer it.

- **Standardization:** The Langstroth type of hives is easily found in the market as they are in standard sizes, and hence acquiring the parts and accessories is not a headache.

Drawbacks

- **Cost:** It is common to find that the first expenses incurred when acquiring a Langstroth hive are relatively high in comparison to other hive configurations, especially when the user has to buy many supers and tools.

- **Maintenance:** Regular updates and repairs are necessary to make the hive healthy, which is a bit tricky for a novice who has many other things to pay attention to.

- **Requires Management Skills:** For one to run a Langstroth hive effectively, they must possess certain management skills, which some new beekeepers may lack.

Warre Hive: A Natural Approach

The Warre hive, introduced by Abbe Emil Warre, adopts a more natural way of keeping bees as the architecture helps honey bees act naturally. This type of hive consists of boxes arranged vertically so that the bees can build the comb naturally.

Benefits

- **Minimal Intervention:** Warre hive's conception is free from frequent hive opening, hence allowing bees to adjust quite comfortably even with substantial monitoring.

- **Better Insulation:** The scheme usually achieves the best ventilation possible, enabling the colonies to avoid excessive heat during the summer and cold during the winter.

- **Simplicity:** The vertical design, along with the few management practices needed, provide an encouraging aspect for beginners, eradicating anxiety pangs from them.

Drawbacks

- **Limited Honey Production:** While honey can still be produced in a Warre hive, some coming of it will not be as much as what is reaped from a Langstroth hive since the management style is less intensive.

- **Complexity of Honey Harvesting:** Honey harvesting from a Warre hive can be somewhat tedious since the comb is not constructed in a detachable manner.

- **Fewer Available Resources:** It could be relatively more difficult to find resources, help, and equipment for the Warre hive as it is not as popular as the Langstroth hive.

Top-Bar Hive: Simple and Sustainable

The Top-Bar hive is another one, which, however, merges an elementary style of construction and beekeeping practice that aims at sustainability. This hive is made of a long rectangular shapped horizontal box with the comb-building ingredients placed in removable bars inside the box at the top.

Benefits

- **Natural Comb Building:** The hive is designed such that the bees are given the chance to build their comb in a natural way, which is believed to be more beneficial for the bees.

- **Easier Honey Harvesting:** Honey harvesting from the hive should be quicker than that of a Langstroth's, as the comb can be removed as a portion of the entire frame instead of the entire frame.

- **Affordability:** Manhattan Top-Bar hives are cheaper to build because they utilize fewer materials than traditional hives.

Drawbacks

- **Limited Space:** The use of horizontal design has obstructions that restrict the number of bees that can fill the hive, and if this is not appropriately taken care of, it can lead to problems.

- **Management Challenges:** It can be less complex in some aspects, but management of the Top Bar hive can still be complex for those lacking carpentry skills, especially in controlling bees to make even comb deposition.

- **Low Honey Return:** Such hives produce less honey compared to Langstroth hives as a result of their smaller and different construction.

Setting Up Your Hive

Location, placement, and hive orientation

After these processes, the next thing that comes after picking your hive type is to ensure that it is properly placed. The success of your beehives can be significantly influenced by the placement and the direction of the hives.

Location

- **Sunlight:** Bees prefer to be in the sun. So, try to orient the hive such that it gets some sun in the morning, as this will warm the bees up and make them readily go out to forage.

- **Wind Protection:** Almost every beehive I have seen is sheltered from the sun and strong wind. Trees and shrubs are natural windbreakers.

- **Accessibility:** Generally, all customers should be able to access the beehives for maintenance or removal of the honey from the hives. You will want to be able to do regular inspections on it with minimal effort.

- **Water Source:** In order for the bees to live, they also require water. Where possible, place the hive near a water source or have a shallow water bowl.

Placement

- **Height:** Ensure that the hives are raised from the ground. House structures, particularly elevation, will prevent water accumulation and pests too.

- **Spacing:** Sometimes separate hives are needed. In that case, provide enough spacing in between the hives to reduce aggression between the colonies.

- **Distance from Neighbors:** Lastly, taking in all factors, distance from other hives is of great importance. It would be better to set a distance not less than 10 feet from any building, and ideally with the entrance directed away from their property.

Hive Orientation

- **Entrance Direction:** The ideal orientation for the entrance of the hive is southeast. This position ideally helps warm up the bees in the morning sun and also encourages foraging.

- **Ventilation:** Be sure to have enough ventilation in your hive. A good air movement assists in maintaining the heat and humidity levels as well as the well-being of the bees.

Hidden Components and How They Work Together

When it comes to the management of bees, it is necessary to know the elements of your hive. Each one is important for the well-being and productivity of your beehive.

- **Hive Body (Brood Chamber):** This space is for the egg but also for the brood, which consists of larvae and pupa. It has parts (frames) where the bees build the comb.

- **Supers:** These upper boxes are on the hive body and are used for storing honey. Depending on the needs of the hive, they can be packed or unpacked.

- **Frames:** Generally, frames contain the wax/plastic-based foundation on which the comb structure is built up by the bees. This makes inspection and honey harvest efficient.

- **Foundation:** It is a thin layer of wax-plated or plastic-coated sheet that is used as a base for building up combs by the bees.

- **Queen Excluder:** It's a mesh that is fitted between sterilizers and trellises to prevent the queen from laying eggs at honey storage blocks.
- **Bottom Board:** It is the lowest portion of the hive, which acts as support and may have an entrance reducer to control the bee entrance.

- **Roof:** The upper part of the hive avoids any bad weather that may be encountered. It should be insulated in order to keep the hive warm during the winter months.

- **Ventilation holes:** They prove useful since they promote movement as well as the regulation of the temperature and moisture content of the hive.

You will also be in a position to handle your hive comprehensively if you appreciate how these components come together. A healthy colony can be maintained with regular checks, and the hive components can be made to perform effectively.

Self-Draining Hives and the Future of Beekeeping

As time goes on and the practices of appending

hives improve, there is an emerging need for intelligent designs, like self-draining hives. A hived design is also applied in this case, whereby moisture is tightly managed, thus being critical for the bees.

Benefits of the Self-Draining Hives:

- **Moisture Reduction:** These hives assist in avoiding the buildup of water, which can invite molds and diseases.

- **Convenience:** This style provides for reduced cleaning and maintenance procedures, thus the beekeepers do not have a lot of work to do.

- **Improved Parameters for the Bees by Maintaining the Right Moisture:** Self-draining hives are instrumental in maintaining the required standard of moisture and therefore create desirable conditions for the bees.

CHAPTER 5

OBTAINING AND INTRODUCING YOUR BEES

The world of beekeeping awaits you! The most exciting phase of your migration to beekeeping is getting hold of your bees and setting them. This section has been made to help you have a smooth transition. It focuses on helping you make the right decisions while also helping you prepare your new buzzing occupants for their new home.

Bringing home new bees to your apiary is quite an excellent feeling, considering that bees are wonderful animals, but such an act comes with a sense of responsibility. Eventually, you can find yourself in that dilemma of going and succumbing to purchasing your bees from a dealer, or you will do it as Mother Nature intended by harvesting them from the wild. All of the options listed above include some pros and cons, and you should take these points into account when making your decision.

Next is the work of getting the bees into the hive. This task is important, as the right introduction might just be the solution towards setting up a productive colony. We are going to tell you down to the smallest detail step in how to install your first colony and make sure that they feel safe in their new house.

Additionally, you will learn how to manage the initial in-hive inspection. Instructions on the beekeeper's ability to notice the signs of a well-functioning hive will inspire him to act swiftly if such problems occur.

From managing an awkward-minded queen to monitoring for different pests, being mentally geared up will smoothen the rough rides of beekeeping given its myriad challenges.

So, let's go into the basics of obtaining and introducing your bees!

Buying vs. Capturing Bees: Pros and Cons

Because it is unrealistic to keep bees if you do not have them, the said decision has to be made first. Two common methods are buying packaged bees or capturing a wild swarm. Each approach comes with its advantages and disadvantages.

Buying Bees

Pros:

- **Quality Control:** When you purchase bees from a reputed supplier, you can expect to obtain good-quality and properly reared bees. Most suppliers provide information about the bee strain and its behavior, and this helps in picking the right variety that is efficient in a certain environment and preferences.

- **Convenience:** Sourcing bees is often quite easy. They can be easily bought by making an online order, and the bees will be sent after a given time, or by visiting an OF discipline beekeeping shop.

- **Fewer Surprises:** Everyone likes bees under this sanguinary sub-heading means working with bees, considering the fact that buying bees can be seen as a plus factor since it brings out all the expectations of the buyer with regard to acquiring the same and reduces the challenges that are associated with the process of capturing a wild swarm.

Cons:

- **Cost:** The necessity of purchasing bees may render an expenditure that is above capturing a swarm. You will have to consider the buying price, transporting costs, and other costs regarding your apparatus.

- **Adjustment Period:** They may require some time to build up in the new bee location. During this time, help them stay healthy.

Capturing Bees

Pros:

- **Cost-effective:** It is possible to avoid buying bees by capturing a wild swarm. Beginning entrepreneurs with limited funds may find this appealing.
- **Local Adaptation:** If you capture local wild bees, they may be good at the local climate and plants, thereby surviving well in that condition.

Cons:

- **Uncertainty:** Activities of capturing a swarm can be very uncertain. You may not be ready for what you are going to find or how healthy the bees would be.

- **Riskier:** Wild bees are possibly more aggressive or dangerous compared with the house bees. This may result in difficult situations for you if you are still a novice.

Conclusion

If you are a novice beekeeper, I recommend buying bees from a reputable seller. If you are a beginner in beekeeping, it will be helpful to start by purchasing bees from a well-known supplier. On the other hand, capturing a swarm can be a thrilling experience if you would like to go through a more dangerous and enjoyable risk. Whatever decision you make, it is wise to learn and gather as much information as possible as it relates to your new hive.

How to Install Your First Colony

After acquiring bees, the next thing to do is to introduce them to their new hive. This process is important because it has the potential to affect how the bees adapt to their new environment. Below is a step-by-step guide on how to properly install your first colony.

Step 1: Prepare Your Hive

This is how setting up the hive looks like, but before you start putting the bees in the hive, you should make sure that the hive is correctly put together. Do these in order:

- **Choose an appropriate location:** Place your hive within a yard that receives sunlight

and protection from heavy winds. Also, low vegetation should be within walking distance and even a body of water.
- **Consider simple access:** Make the hive simple to reach for checking and upkeep.

- **Components:** Make sure all box parts are placed; task frames, layout, and queen exclusion are actually dependent on use.

Step 2: Introducing the Bees

- **Timing:** Weekends are suggested to place the bees without much movement and prevent offending them. This step should be done no later than afternoon when it is not too hot so that the chances of bees getting agitated are minimized.

- **Prepare Your Equipment:** Get safety gear and have your tools all neat and ready to go. A smoker, a hive tool, and a package of bees would be required for this task.

- **Opening the Bee Package:** This procedure should be done with extreme caution, mostly in regards to the spillage of bees.
  Remove the cover of the bee package with care.

- Find the queen cage since her majesty is always within a small piece of transportation gown.
- **Installing the Queen:** The queen cage is placed in the middle of the hive on the frames at the center.
  The bees will slowly chew or eat the candy plug in the cage in which the queen has been enclosed, and hence, the queen is released and the process of accepting her starts.

- **Introducing the Bees:** After doing that, simply shake the bees or pour the bees right in the hive where they expect to settle around the queen.
  Tightly seal the hive so that the opening remains unblocked.

Step 3: Monitoring Hive health and activity

After installing your bees, it's important to monitor their activity:

- **Search for Activity:** Check out the hive entrances for foraging bees that are returning with pollen as horizontal surface activity.

- **Settle Queens:** Around three weeks after adding colonies in the new location, it would be sensible to check whether the queen is out and laying, but combs are already being constructed and stabilized.

What happens on the first inspection?

It is a very exciting moment owing to the revision of the first hive during the analysis. Although, to some extent, one can also become more anxious.

This is what you need to have in readiness:

Recognizing the Signs of a Healthy Hive

- Noble Activity: Active hives will have bees flitting in and out. Find pollen and nectar-returning foragers.

- Frame with Brood: It contains eggs, larvae, and even pupa stages of immature insects developed in hives. When you dismantle the frames from the hive, let's check the brood frames. A consistent building activity means a good layer of the queen.

- Honey, which has been sealed Look at the frames that are loaded with honey that has been sealed up. This sign shows that the bees are preparing food reserves.

How to Address Common Challenges

- **If You Don't See the Queen:** If You Don't See the Queen: In some circumstances, the queen may prove elusive. What should be worried about instead of adults is brood pattern and signs of activity.

- **Aggressive Behavior:** If the bees show themselves as overaggressive, then close the hive and do not enter it for the time being. When it will be calmer, the inspection should be repeated. This can occur due to weather conditions, such as heat or due to agitation due to a new bee colony.

- **Signs of Disease:** It is also important to pay attention to any classic signs of disease that show up in the bees' behavior or their physical appearance. Seek professional medical advice if symptoms of disease are observed.

CHAPTER 6

SEASONAL HIVE MANAGEMENT

Beekeeping is one of the most assuming activities that requires a corresponding outlook to the varying seasons. Each season has its own relevant issues and opportunities, and knowing how to deal with the bee setup during these periods is important in maintaining the population of the bees and the activity of the hive. In this address, we'll walk you through the procedures to undertake the management of your hives in the spring, summer, autumn, and winter to help you preserve healthy colonies throughout the year.

While doing beehive management at irregular intervals, the aim is not merely the survival of the bee colonies but the headiness of the bees in work—honey production, pollination, and preparations for the winter. Girard is the third-generation beekeeper who is very active in the practice and believes that knowledge is power and timing, and knows the physiology of the hive very well in order to get the right actions at the right time. Now let's move forward with identifying the seasonal apiary management activities and their significance to the productivity of bees in all seasons.

Spring: Preparing for Growth

Starting with spring, which is arguably the most active period for all beekeepers. As the snow melts and the temperatures get hotter and the blossoms start blooming, the life of the colony begins to rouse. It is that time, which is characterized by developing and maturing, therefore crucial in taking care of the hive and preventing swarming.

Hive Health

Spring is the season when your bees are replenishing their population. After such a long period without laying many eggs, the queen very much increases her egg production, and the worker bees are active in searching for and storing pollen and nectar. But even then, it should be remembered that before the population of the colony is in excess, it is crucial to carry out intricate evaluations.

Steps for Spring Hive Inspection:
- **Through the eyes of the queen and the brood, check the condition of the brood frames:** On the first spring inspection, the

most important aspect is that the queen is confined to when and how many eggs she is laying and whether or not the brood pattern issued has healthy, smooth growth. Perform a diagnosis of brood diseases, like foulbrood or chalkbrood, on a sample basis.

- **Evaluate the presence of pests:** Look out for some of the pests, including varroa mites, as they should be kept at reasonable levels. This season is also appropriate for mite treatments, and the rush for honey begins.

- Furnish your bees with nourishment when warranted. In case of prolonged storage of a limited winter ration, thrifty bees should be fed sugar syrup to plus up the colony until the blooms come forth.

Swarm Prevention

Swarming can be defined simply as a process through which some of the subjects of a colony's hive move out, leaving the rest of the bees behind to form a new colony. For a beekeeper, such a process can render loss by way of half the bees in the hive. Therefore, swarm prevention measures should be taken at the very least during the spring.

Tips for Swarm Prevention:

- **Raise sufficient space:** Where the hive population of the colony expands, there is a need to add supers or extra partitions to the hive so that the bees have enough room to grow in numbers.

- **Observe what happens to the swarm cells:** In focus are the swarm cells, which are a sign of the in-hive brood or the frame orientation cells. The cells are called brood frames; those large queen cells are preparations of the hive to swarm. If some of these cells are available, then it may be ripe to biopsy the colony.

- **Hive ventilation:** With the increase in temperature, bees are likely to suffer from high-temperature stress. Adequate ventilation inside the hive alleviates the strain on the colony, which reduces the chances of swarming.

Summer: Managing expansion and honey production

In the summer, your bees will be the most active. When the colony reaches its optimum size, bees become industrious in the collection of nectar and the manufacture of honey. But summer also has its challenges, such as the health of the colony and honey-making.

Harvesting Honey

Around midsummer, you can expect your bees to start capping the honey, which means it is now ripe for harvesting. The focus here is to ensure that honey harvesting is done without upsetting the equilibrium in the beehive.

Steps for Harvesting Honey:

Capped honey is present when bees have filled their cells and sealed them with wax. Honey that has had some nectar added should not be sold. Only deal with fully capped honey in order to prevent dealing with wet honey.

Centrifugal honey extractors are' damage-free' for comb honey and the most effective in governance for most improvisers of bee honey. For the majority of beekeepers, it is one of the most effective ways to use a centrifugal honey extractor, and it is particularly suitable for removing honey from honeycomb frames without causing damage to the hive's structural combustible.

Cordon to achieve maximum honey removal: Although it is not desirable to leave the colonies empty when harvesting honey, the honey reserves of the colonies should be maintained when harvesting so that the remaining amount of honey is about 40–60 pounds so that the bees can cope with the winter.

Maintaining Hive Health

Above all, summer can also be a stressful time for your bees because of adverse conditions such as drought or high temperatures. However, proper hive management will help ensure that your colony remains in excellent health throughout this strained period.

Summer Hive Health Maintenance Practices:

- **Promote good airflow:** August continues to be associated with high temperatures, and as such, overheating is likely to be a challenge; hence, we must ensure that the hive is adequately ventilated. Some beekeepers slightly tilt the roof of the beehive to improve interior conditions, while others use a screened bottom board to enhance airflow for the bees.

- **Scout for pests:** It is also at this time that these means show their effectiveness in controlling pests like varroa mites, wax moths, small hive beetles, etc. Bimonthly hive checks are essential, and pest control measures are advisable.

- Get a shallow basin of clean water. Workers also require water to aid in cooling the hive and thinning of honey. Provide an incomprehensible low basin of clear, calm water, which is okay because the bees may drown.

Fall: Preparing for Winter

As summer ends and fall begins, it's time to help your bees get ready for the winter. Fall is focused on feeding bees enough food stores and proper insulation for optimal overwintering.

Nourish Your Bees

In autumn, as the weather cools down, the explanation of bee hibernation activity comes into play; the bees' work and activity is gradually reduced. However, sufficient food stores will still be needed in order to survive wintertime inactivity. If your bees have not gathered enough nectar, you might have to consider supplementary feeding.

Types of Fall Feeding:

- **Sugars syrup:** A common synergetic mixture for Acomb bees or bees in general syrup mainly comprising sugars. It is 2 parts of white sugar to one part of the water.

- **Fondant:** In colder regions, solid sugar fondant, commonly known as fondant, can be used as a last-minute feeding solution to keep bees alive until the weather improves.

- **Pollen patties:** Supplementation with pollen substitutes is also necessary to ensure that the health of your bees is maintained during the months heading into winter.

Insulating Your Hive

Insulating Your Hive Proper preparation of your hive for winter includes preparing it to the external cold conditions. Lower temperatures and dampness are dangerous for the bees, and hence proper insulation is crucial.

Tips for Insulating Your Hive:

- **Add an inner cover:** Inner covers should also be considered to provide insulation to brood areas in cold months to prevent excessive heat loss from the hive. Extra inner cover helps to prevent moisture condensation across the screens and helps prevent the molding of the hive surface.

- Use a moisture board. Cold is more benign to bees compared to condensation covers and quilts. A moisture board or quilt box is a cover that is placed over the frames or crown board of the hive to absorb excess moisture, which may eventually flow back on the bees.

- **Reduce the entrance:** This helps to cut back on the amount of cold air that would have rushed in through the entrance, and in addition, that would barrier pests such as mice, strange rodents, and the like from accessing the hive.

Winter: Overwintering Your Colony

Both bees and beekeepers face the toughest seasons in winter. In this season, bees remain mush together for warmth, and the activities in the hive, if there are any at all, come to a standstill. As a beekeeper, one of your jobs is to ensure that your colony is stocked up sufficiently to last through the long cold months ahead.

Keeping your hives warm and safe

Bees are extremely tough, but not to the extent that they will not require assistance during the winter period.

Winter Survival Tips:

- **Ensure food supplies are checked:** This does not imply that you should always be opening your hives during winter, but once in a while, it is necessary to make sure your bees have enough food. When it comes to food supplies, sugar cakes or fondant, for example, are recommended in case all sources have been exhausted.

- **Avoid excess moisture:** Insulation wraps are very useful during winter; however, moisture that gets trapped in can also be very detrimental as it can condense and freeze on the bees.

- Protect the hive from wind and predators. For example, you might want to consider covering your hive or having wind stoppers to block strong winds during the winter. Also, filters should be used so that mice and other small creatures cannot gain access to the hives.

Keeping track of hive activity

Throughout the last months of the year, your bees still keep on performing their functions. They actively keep the brood warm and attend to the queen. You know that you will not observe many movements outside the hive; however, I think it is good practice to keep checking now and again.

Important Things to Note During the winter

- **Hear the sound of bees:** There is a safe sound of buzzing when all the bees clustering will be packed within the hive. Any such instances of silence would ring alarm bells.
- **Optional using a thermal imager:** It is the practice of some beekeepers to use thermal imagers to stalk the location of the cluster without having to open the bees and interfere with them.

CHAPTER 7

HARVESTING HONEY AND OTHER HIVE PRODUCTS

Harvesting honey and other hive products is perhaps one of the most fulfilling and productive aspects of keeping bees or beekeeping. Are you an allergist? After such a long period of care, nurturing, and taking care of hives, you at last taste the fruits, or, to be more precise, the sweetness of your hard work. Now, honey, is just the start. It is quite possible that several valuable by-products, such as beeswax, propolis, or royal jelly having their own practical and therapeutic applications can be harvested from a properly cared for hive.

Introduction: The Sweet Rewards of Beekeeping

This section cover the techniques to remove honey from the frames that allow for the use of most experienced and new beekeepers alike. We'll also discuss other hive products and how to use them, including insider harvesting tips. If you are in it just for the love of honey or art from natural materials or want to use the health-enhancing properties of royal jelly and propolis, there is definitely something to look for in why understanding every aspect of the hive is important to a beekeeper.

How to Harvest Honey the Right Way

Honey harvesting has to be and is one of the most appealing aspects of keeping bees, but there need to be some techniques so that the health and productivity of the bees are maximized while getting the most out of one's hives. It is not only a yummy substance that people enjoy, but rather the basic food of your bees. So it is essential to take care of the need for harvesting any honey, but in such a way as to leave enough for the colony, especially through the winter season.

Extraction techniques for every beekeeper

A) Traditional honey harvesting

Honey is extracted from the comb with the most common means of honey harvesters a honey extractor. A honey extractor is a device that turns the hive's honeycomb frame, thus extracting the honey. A step-by-step guide for each stage of such processing will be given below:

- **Check the Frames:** First, check whether the honey is now ready for harvesting. Honey is considered ready when at least 80% of the honeycomb cells where honey is stored are covered (capped) with wax sheets. It assures that the moisture is low and the honey will be a good stash.

- **Prepare the Hive:** Tenderly puff some smoke into the hive to quieten the bees before the frames are pulled out. Use a battery-operated bee brush or similar gentle sweeper to sweep off the bees from the frames that are being harvested.

- **Uncapping the Honeycomb:** Remove the wax caps from the honeycomb cells using an uncapping knife or any suitable implement, such as a dishing fork. Care should be taken not to injure the honeycomb structure.

- **Place in the extractor:** Attach the frames to the honey extractor. The honey extractor operates on a centrifugal force to eventually draw honey from the frames and direct the honey down in the extractor.

- **Filter the Honey:** After the honey has been drawn out, it must be passed over a mesh screen to get rid of the wax, bee particles, and other materials. Thus, cleaning and sanitizing of the honey and preparing it for ready storage or sale are achieved.

- **Storing sterilized containers:** After the honey is bottle-honey is ready for bottling, it must be packaged in a clean and heat-sealed container and aged. Honey can be stored for a long period of time if it is kept in a cool, dry place.

B) Crush and Strain Method (No Extractor)

The crush and strain method, on the other hand, is one of the simplest ways of extracting honey that does not require a honey extractor. Because it eliminates the use of the honeycomb, it is an ideal way for novices to gather honey without expensive tools.

- **Take Out the Frame:** Slowly detach the frame of the honeycomb from the super.

- **Make a Pulp out of Honeycomb:** Take a clean knife or spoon and mash the honeycomb into a bowl to get the honey out.

- **Filter out the Honey:** Take the mixture of honey and crushed honeycomb and pour it

through a fine sieve or strainer to remove the pieces of wax.
- **Honey Preservation:** In the end, after all the operations are completed, the honey is strained and put into jars for storage.

C) Comb Honey Harvesting

The harvesting of honey can be done in one natural form for other beekeepers: in the honeycomb. Comb honey is appreciated for its quality and sug-eating activities of people. To collect comb honey, one just needs to harvest pieces of honeycomb frame and bottle it. This method is not so popular nowadays. Still, it has a certain liking among honey lovers.

The Uses of Beeswax, Propolis, and Royal Jelly

Other than honeybees, other interesting products can be harvested from the bees apart from this plentiful quantity of honey that has been put into use since time immemorial.

Here's a summary of these special substances that have been put to various uses while working with the bees

- Rendering Beeswax for Use in Candle Making and Crafts

Beeswax is secreted by worker honey bees and used to form the hexagonal structures of honeycombs in the hives. After harvesting, beeswax can be purified and used as follows:

- **Making Candles:** The reason as to why beeswax candles are embraced is because they last a long, burning time, are free of toxins, and have an have an appreciable aroma. To make wax shaped into the desired forms, candles, or insert an appropriate Bee's pure, unadulterated, edible go beside the wick in the center of the cavity, which is mostly the burnt-out stick.

- **Lip Balms/Lotions:** Many cosmetic preparations, such as lip balms and lotions, possess a nature of beeswax properties. It serves to seal in moisture and protect skin.

- Polishes and Coatings: Beeswax can be used as a natural polish for wood and leather, providing a protective and shiny coating.

Royal Jelly and Propolis: Nature's Healing Remedies

Royal Jelly: This is a substance secreted by the workers and fed to the queen only. Royal jelly is prized for its high levels of vitamins and proteins.

Royal jelly is also thought to be an energy enhancer and a supporter of the immune system, and consequently, a beneficial dietary supplement.

Propolis: Propolis is a sticky, resinous substance that insects make from plant material such as tree buds, leaving the sap flows. It is also used in the hive, a building more like construction sealant, which has antibacterial and antifungal activities.

Propolis can also be made into anointment types that are good for wiping away swellings about a wound.

Tips for Processing and Storing Hive Products

- **Beeswax:** Beeswax comes from the bee and can be molded into different things. Always heat it carefully under the flame and do not scorch the wax. A wax-dedicated pot, or double boiler, is what is used to heat the wax because, heated instantaneously, it may burn. Strain while the wax is still hot, and all lumps will be removed through ordinary sieving.

- **Royal Jelly:** Emphasis should be placed on the fact that royal jelly should be harvested with great emphasis. The honey makings are located mostly in the queen cells, cells that had been secreted for feeding of the available would-be queens. The use of a refrigerator is for the probiotic content instead of the UV light of the summer sun.

- **Propolis:** It is collected by scraping it off the surfaces of the bee nest with a scraper. After that, one can dissolve the substance in alcohol in order to prepare the tense of it or heat it to form balms.

Tips from Expert Beekeepers: Maximizing Your Harvest

Beekeepers who have had enough experience will tell you that the efficient and versatile harvesting of hive products is critical for the growth and sustainability of the apiaries. Here are some of the variations your expert suggests you regard while harvesting:

- **Plan Your Harvest According To The Annual Days:** Honey bees are extremely hard and happiest working in the late spring and summer seasons when there are lots of flowers. Keep this in mind while planning your harvesting period, as that is when you are going to get the maximum honey out of your   bees.

- **Do not harvest everything. Do Not Starve the Bees:** Do not be greedy and harvest excessively. There should always be enough honey to feed the bees, and even more so, especially during cold seasons. Make it a rule, for instance, to leave behind a minimum of 30 to 50 pounds of honey in each hive.

- **Always Monitor the Hive:** An old saying goes that there is health in wealth, and so is a hive. You need to carry out regular hive inspections to make sure all your bees are content and free from pests, diseases, and all other factors causing them stress. We would like our customers' goals and expectations to translate into higher and better-quality honey and hive products from our clients' bees.

- **Have the Right Tools:** With the right tools, like extractors, strainers, and protective gear, you will not only make completing the work easier but even enhance productivity during the harvesting process.

- **Look for More Than Honey:** Do not just harvest honey. Try other hive products like beeswax, propolis, and royal jelly. Each product boasts some usefulness that can enhance your beekeeping.

CHAPTER 8

BEE HEALTH AND HIVE MANAGEMENT

Every beekeeper knows that it lies on him or her to ensure the health of the bees, which happens to be one of the essential duties that a beekeeper must undertake. Healthy bees translate to productive bees and, more crucially, the ability to endure whatever the challenges encountered, be it disease, pest, or environmental challenges. In this chapter, we will take a step-by-step approach as concerned with the ways you can use to make sure that your colony is as healthy as can be and does not weaken. We'll review the most common diseases and pests that target your bees, the preventive measures and management approaches that you can consider to control such agents, and how you can foster the health of the hive through proper feeding and management.

As either an aspiring or indeed an actual beekeeper, being savvy enough to spot the signs associated with emerging bee health problems and, just as importantly, what to do about them, especially when it comes to honey bees, will not only safeguard your bees but also make sure that your hives exist and flourish for a long time to come.

Recognizing and treating common bee diseases

As any other living organism, honey bees may also succumb to many ailments or diseases. It is worth trying to understand and solve such problems to prevent the spread of the same disease between colonies or within a colony.

Varroa Mites

Whereas Lotka and Volterra used insect populations like these in order to die out, Varroa destructor, better known as Varroa, is a pest that is perhaps the most well-known within beekeeping. The little reddish-brown parasites jump on the adult bees and developing larvae and suck the bodily fluids out of them, thereby weakening the immune systems of the infected. Varroa mites don't just kill the bees; they also introduce viral diseases that drive entire colonies to collapse.

Indications of a varroa infestation include the following:

- Seeing bees with visible mites upon opening the hive.

- Young immature bees with poorly formed legs or wings.
- Lesser quantity of late stage larvae or erratic distribution of late stage larvae.

In the case of Varroa technology, techniques include:

- **Chemical treatments:** There are therefore numerous chemical therapies, and their application against Varroa mites is quite common. It is hence very important to adhere to the instructions and practice them precisely in order not to hurt the bees and pollute the honey. The common ones that have been used include oxide acid, formic acid, and amitraz.

- **Biological controls:** Some beekeepers apply biological control using organisms that are known to feed on pests or chemicals that kill mites but are not harsh on the bees. One of such techniques is called powder sugar application, which makes bees clean their bodies and remove mites.

- **Mechanical controls:** Trapping drone brood is a method that uses drone brood frames, which are preferred by miters to reduce their number in the beehive.

Adhering to this principle will help in managing Varroa mites disease successfully. Integrated pest management (IPM) trades are often the best because they keep checking for the presence of each pathogen while trying to eradicate them using multiple strategies.

Foulbrood

Foulbrood There are two main varieties of foulbrood. These include American foulbrood (AFB) and European foulbrood (EFB). They are both bacterial infections of the larvae, and if left untreated, will result in colony loss.

American foul brood is the more severe case and is caused by the bacteria, Paenibacillus larvae. The cut infected larvae go brown and dry and fill the cell with a sticky substance and bad smells.

European foul brood is a disease caused by Melissococcus plutonius and does not display as much aggression as AFB does. Larvae that are infected usually die before they are sealed inside their cells, and the infection seems to self-resolve and does so quite often in hives with strong colonies.

Treatment options for foulbrood:

Interpetition AFB and Epanes oralantimids. Binning gently, chronologically, woefully affected hives:

Fearing AFB will literally give a death sentence to disease-bearing, reused frames and nooks of brood and soon-to-be bees of the afflicted colony to contain the outburst into any fraternity.

Antibiotics: EFB is at times treated with antibiotics such as oxytetracycline as a last resort. There is no way to control EFB once colonies get sick with it because there are no effective treatments apart from nursing the bees to strengthen them when they are under attack.

There has to be thorough monitoring of hive populations, restriction of movement of hives and follow practices that enhance good hygiene and parasite control, and elimination of infected colonies to curb foulbrood spread.

Nosema

Nosema is a honeybee disease caused by a fungus that invades the mature bees' digestive system and compromises honey production as well as the strength of the colony. There are two forms of Nosema, two types, and that is Nosema apis and Nosema ceranae, but it is the latter that has seen increased prevalence of late.

Some indications of a cone nosema infection include:

- Bees are either disoriented or too sluggish.

- Fecal stains near the hive entrance.

- Less honey and fewer productive hives.

Nosema Treatment:

- **Administering Fumagillin:** This antibiotic is used to treat a nosema infection mainly but must be administered with caution because it poses a threat to honey contamination.

- **Preventive measures:** Make sure your bees are well fed, especially in the early spring and late autumn, before they get stuck in Nosema. Also, if the hives are not soiled and have ample ventilation, their chances of outbreak are less.

Managing hive pests and preventing infestations

Managing hive pests and preventing infestations it's not only diseases that affect the honeybee colonies, but also there are many pests within the colonies that can damage their health status and the output. Therefore, it is very vital for every beekeeper to equip himself with pest management skills.

Hive Beetles

Small hive beetles are defined as very small black insects that either infest hives or lay eggs and kill the comb and honey reserves. Warm and humid conditions favor their growth, and they will easily devastate a weak colony.

Indicators of a hive beetle infestation:

The beetles are seen moving in and around the bees and the combs.

Bumpy combs and a pungent odor of leavened beeswax comb inside the beehive.

Prevention and management

- **Trap placement:** For instance, beetle traps that are incorporated externally on the hive and also such baits such as oil and diatomaceous earth incorporated in the traps have been found to be efficient in managing the beetle population.

- **Strong colonies:** The hive beetles can easily be managed by ensuring the colony established is a strong one. The strong colonies will in fact drive the beetles outside the beetles wax and also the hive.

Wax Moths

Wax moths are another pest that can invade hives of bees, mostly weak ones. They cleave wooden comb material and lay their larvae, waiting for them to hatch. The caterpillars of the moth will eat a lot of the wax, which forms silk and spins webs that are very destructive.

Prevention and management:

- Do not store empty frames out of the humid where the empty infesting myths are most likely to oviposit.

- Limit exposure of all hives with moths or their larvae to other hives to reduce or prevent contamination.

Integrated Pest Management Strategy (IPM)

IPM is an approach that combines biological, mechanical, and chemical pest control methods that are done with low risk to bees or the environment.

However, good features can be found in integrating several pest control strategies because it most likely reduces the possibility of pest resistance to one method. Also, the impact on the hive will be reduced.

Key components of IPM:

- Routine monitoring of hives and detection of pests in the early stages.

- Biological agents, in such instances where breeding of new biological agents helps to eradicate the problem.

- Device controls, including traps or physical barriers

- Chemical agents are utilized only after other methods have been employed and in a judicious manner.

Maintaining healthy bees through nutrition and care

It's hard to overstate the value of sound nutrition in a bee colony's health. To aid in their development, the bees have to be given many kinds of pollen and nectar in order to supply the vitamins, minerals, and proteins their bodies need.

Supplementary Feeding of Bees

There are times when natural forage may be difficult to obtain, such as in very early spring or in very late fall; at such times, one may have to do something by way of supplementary feeding.

Options of feeding:

- **Sugar syrup:** One of the applications of sugar syrup in feeding bees is in the spring, where a 1:1 mixture of water and sugar is used so as to stimulate brood rearing. Later in the season (in the fall), a thick syrup (2:1) is used, which helps bees store more food for the winter.

- **Pollen patties:** These are pollen substitutes that may be fed to the bees at non-pollen foraging seasons.

However, it is nearly advisable to ever feed bees on honey from outside sources. The reason being the honey may introduce foulbrood and many other diseases to the hive.

Water and hydration

In hot weather, bees require a steady supply of clear water. Therefore, shallow dishes filled with water should be placed in front of the hives and filled with small stones or sticks as a support for the bees when they drink.

Hive insulation and protection.

- Your colony's winter survival depends on proper hive insulation. Bees cluster together to keep warm, and as strong as they are, insulation offers them protection and less use of energy.

- **Wrapping your hives:** Wrap your hives with insulative materials in autumn, but make sure that the insulation is breathable to prevent condensing moisture.

- **Windbreaks:** Siting the hive close to some natural windbreaks or artificial ones will also reduce the windchill factors outside the hive and also prevent cold draft under the lids.

CHAPTER 9

THE ROLE OF BEES IN POLLINATION AND BIODIVERSITY

Among all living organisms, the honeybee has incomparably marked itself as being possibly lifesaving pollinators. They reproduce the hosts of organic species and balance many ecosystems. Apart from honey production, bees are essential as they assist in the process of pollination, thereby enhancing the variety in living things. Contemplating the growing environmental issues, it has become more important to appreciate the fact that bees are incredibly important to the wellbeing of both the environment and agriculture. It is wonderful to think of this equilibrium in which beekeepers can have a positive influence, and in this chapter, we will correct this imbalance and show you, them, and all of us how beakers increase pollination and biodiversity, how you can help them, and how this influences the global ecosystem.

Why bees are critical pollinators

Bees are also imperative to the survival of many plants, animals, and, most especially, humanity, due to their role as pollinators. Pollination is defined as the movement of pollen from the male anthers of a flower to the stigma of the same or another flower to cause reproductive fertilization and the generation of seeds and fruits. In the absence of bees, most plants would have difficulty flowering and hence threaten the food chain that depends on them.

The Global Impact of Pollination

According to the studies of various scientists, bees are contributors to the pollination of around 75% of the flowering plants present in the world. These plants and others, which include many food crops that support human life, are the building blocks of natural ecosystems. Some fruits and vegetables, including apples, almonds, berries, and cucumbers, among many others, are dependent on bees for cross-pollination.

There are other primary indirect contributions to bees, such as the provision of meat, milk, and other livestock products. Many animals consume plants that are bee-pollinated in their feed, and therefore it is certain there would be a disruption not only to that system but to the entire food production system without bees. It is overreaching, however, to say that the bees' contribution to humankind is limited only to food production. They help in conserving the plants, thus helping animals of various kinds with their food. This interdependence again relates back to how important the bees are to food production and nature.

What Makes Bees One of the Best Pollinators

Bees display certain traits that aid entomophily. As they fly into flowers seeking nectar, pollen gets caught on the fuzz, which decorates their exoskeletons. A bee commonly engages in drifting about through multiple patches of flowers, and therefore, an additional gain in cross-pollination is achieved. Pollination requires loyalty on the part of the pollinator, which bees are able to offer too. Once a beneficial flower has been found by the bee, the bee visits every flower of that type until enough of the correct pollen species is received for fertilization of the flowers.

Long flying distances and visits to hundreds of flowers each day make bees the unsung soldiers in farming and the preservation of biodiversity. Not only is this practice beneficial for increasing harvests, it also helps in preserving the plant variety that many ecosystems rely on.

Using Your Bees for Pollination Services

While preserving your colony, beekeepers improve the environment. Beekeepers may hire out their bees for use in pollination, which enhances agriculture activities while taking care of the ecosystem.

Renting Your Bees for Pollination Services

There are various farmers who hire commercial beekeepers to pollinate their crops with the use of their bees. This is a win-win arrangement: the higher the farmer's crop yield, the more the beekeepers earn additional financial benefits from their colonies. Typical crops that depend on bees and thus pollination contracts include almonds, apples, and blueberries.

With this excitement about providing pollination services comes the obligation of preparing the bees for the role. It is imperative to step up your game by ensuring adequate colony health since weak bees will lack efficiency as pollinators. Ensure that there are no hard and rough procedures when the bees are being moved from the planes to the hives or from the hives to the farms. To calm the bees while transporting, it is possible to use the technique of ventilating and covering them with screens so that there is minimal stress-causing movement.

The advantages of the pollination services are not limited to the farmer or the beekeeper but also nurture the environmental conditions. It has been demonstrated that as they move, beehives can also help with the reproduction of the indigenous wildlife and, in turn, stimulate the biodiversity of the area.

Maximizing Pollinator Wellness

When investing in apiculture and expecting the bees to pollinate a colony, it is important to focus on the bees' health. It includes regular events meant for CLI checking of the hives, feeding the bees, and keeping them away from pesticides as well as pests. They tend to experience misuse of pesticides or lack of proper management and therefore end up being weak and not being effective in pollination.

One aspect of promoting the health of the pollinator, in this case the honey bees, is maximizing the forage available to them. A lot of crops grown fall into the category of monocultures, which offer poor nutrition. Beekeepers, however, can grow an array of flowers free from pesticides around the hives to enhance the nutrition of the bees. Adequate nutrition reduces diseases; therefore, through diet, poor pallor performance is taken care of, hence excellent pollination and management of crops in their phase of growth.

Creating a pollinator-friendly garden

Every garden lover, be it a beekeeper with an extensive collection of hives or just an individual enthralled by gardening, can make a difference in the preservation of pollinators through establishing a pollinator-friendly garden. This type of garden does not simply ensure that your honey bees are happy, but it also caters to many other wild pollinators, such as turtles.

How to Select Suitable Plants and Flowers

When it comes to establishing a pollinator garden, one of the most fundamental factors is the selection of plants. Different types of bees or other pollinators will be drawn to different kinds of flowers, so you need a variety. Native flowers are a good idea, as they are often favored by local bee species. Complete this as well by selecting plants that flower at various intervals over the growing season to make sure that food will always be available for your bees.

Some flowers known for attracting bees include:

- Lavender
- Sunflowers
- Clover
- Marigolds
- Wildflowers (such as Black-eyed Susans and Coneflowers)

- Borage
- Rosemary

Deciduous and flowering trees can also be included in the designs to attract pollinators, as they provide budding pollinator seekers both food and cover. Trees such as these apple and cherry trees are quite popular springs as they have a lot of nectar bearing. In the autumn, there are also charm bushes such as hawthorn and holly that provide food.

Beekeeping and conserving biodiversity are synonyms

Beekeeping is also more than just beekeeping for honey. If you construct and take care of your hives properly and make the surroundings suitable for all kinds of pollinators, you will create conditions favorable for conservation. Bees and their fellow beneficiaries play a crucial role in every ecosystem. So, when you enhance pollination, not only does it lead to better harvest production, but also the population of natural flora and fauna that depend upon those plants is improved too.

While you are trying to help biodiversity in your ways, when it comes to maintaining your garden or growing flowers, do not use any harmful sprays like pesticides or herbicides. Such substances may poison the bees and other pollinators, ruining the aim you are trying to achieve. Attempt the use of either chemical-free or naturally organic ones, and do not forget about the companion planting techniques that effectively work against pests.

Illustrative Case Studies

Numerous efforts have been taken on a global scale to foster the concept of pollination and conserve biodiversity. Initiatives like "Bee City" in Canada and "The Pollinator Partnership" in the US urge cities and communities to build habitats that attract these insects. These efforts also help in educating people on the necessity of bees and making them act within their regions.

Concerning this aspect on a smaller scale, individual beekeepers and gardeners may also be involved in these activities by creating biospheres that enhance biodiversity. Whether it involves urban beekeeping projects, maintenance of school pollinator gardens, or use of organic farming techniques, all these efforts work towards the improvement of biodiversity and the survival of other vegetative and animal life forms.

Little Things that Readers Can Practice

A practice of keeping bees or providing care within a garden, these are the things you can positively do in order to enhance the sporting of bees and their related functions, i.e., pollination, not forgetting biodiversity:

- **Engagement in multiple flowering pollinators:** The entire flowering period ranges from early spring to late days of fall.

- **Use of pesticides:** Rather use pest repellants that are organic or bring in biological control.

- **Make water available:** Bees, just like every other living thing, need clean water for hydration. Put a deep plate with some stones or a small pond where bees can go drink safely.

- **Make spaces for bees:** Put in place bee colonies in the backwoods, build bee hotels, or leave regions of your yard free from cultivation for stray ground bees.

- **Help out with projects on pollinators:** Get engaged with local ad hives, lend your hand to urban agriculture initiatives, or back any local campaign that seeks to make spaces loved by pollinators.

CHAPTER 10

TROUBLESHOOTING COMMON BEEKEEPING ISSUES

Beekeeping has its own challenges, even if it is rewarding. From swarming bees to aggressive colonies, these problems can sometimes feel overwhelming, especially to the novice beekeepers. It is true that certain factors can make a part of the beekeepers' work disappointing, but most beekeeping problems, if dealt with with knowledge and adequate calm focus in looking for solutions, are easily dealt with. In this chapter, you will learn and understand some of the problems beekeepers usually encounter, as well as the smoke and mirrors that can be employed to fix the problems. You will also learn how to avoid swarming, handle the aggressive bees, and deal with the absence of a queen bee and stressors in the environment. By the end of this chapter, you shall have gained the skills and the feeling of how to run a colony of bees in a way that encourages productivity and health even when one is faced with problems.

Swarming: How to Prevent and Manage It

Swarming is considered a natural process among the lifecycle stages of a bee colony; however, it raises concerns when it comes to the beekeepers. When sourcing new dwellings, more than half a hive, including the queen, will abandon the colony at once to do so. This disruption enables the acquisition of bees as well as affects the level of honey produced, which causes most beekeepers to want to prevent or manage swarming.

Why Do Bees Swarm and How to Stop This Behavior

Swarming is a reproduction behavior in bees that occurs when the colony gets overcrowded, the temperature in the hive becomes high, or the bees think they need more room. It's also an effective means of ensuring the propagation of their species. However, while it also indicates that the hive is strong and healthy, one will lose the bees from their BC, and you will be running around looking for ways to prevent it.

Your best options to control swarming include:

- **Provide Adequate Space:** having enough room within the hive for the expansion of colonies. The addition of supers (the boxes that hold honey) enables the bees to be less crowded, minimizing the chances of crowding.

- **Surveillance of Queen Cells:** These are the preventive methods to consider when controlling swarming. A queen cell indicates to the beekeeper that the colony is ready to swarm. Therefore, regular inspection of hives may assist in identifying the presence of frame-capped queen cells before swarming. These may include splitting the bee colony or re-queening the hive.

- **Splitting the Hive:** This is simpy where, for instance, if your colony is able to thrive well and grows very large, then reallocating the existing colony into two colonies may be worth it. It simulates the natural phenomenon in which some bees will leave the main colony with a swarming queen and head out with some worker bees.

- **Re-queening:** It is likely that swarming will be increased where there are old queens. This can help prevent swarming since the queen is replaced every one or two years.

When the rearing of honeybees in a beehive is concerned, if the bees start swarming, there is no cause for alarm because the swarm can still be caught and boxed in another hive that has been prepared for use. It is particularly simple to catch swarms since they are calm in nature and therefore easier to place back to the molten prepared comb casts.

Dealing with Aggressive Bees

From time to time, you may have to handle more aggressive bees than you have already trained. The hive inspections as well as retrieval of honey from the hives may be quite challenging and stressful in these cases. It is important, in beekeeping, to comprehend the reasons for aggressive behavior in bees and the ways of restraining it.

Identifying the causes of aggression

- Honey bees can become aggressive if the bees sense that their hive is under attack. There are a number of possible explanations for this behavior.

- **Queen Problems:** A whipping or missing queen may alter the colony being more defensive. There will be turmoil in a beehive until the queen is reared and mounted on a throne and sworn in as the new queen.

- **Weather Conditions:** Clouds or thunderstorms can be considerable irritants to a bee colony. It is advisable to refrain from opening hives during such periods.

- **Pests and Predators:** During the summer, if released, wasps or hornets will chase the bees back into the hive, and bees will be on high alert and will be stingier.

- **Improper Handling:** Mismanagement of the hives may also break down the colonies, making the bees hostile. Thrashing about or moving quickly may also add pressure.

Dealing with and Soothing Bees in a Non-Threatening Way

Here are some ideas to deal with aggressive bees and protect yourself and the colony:

- **Do the Work When Resting On Reverse Days:** When the weather is nice and the sun is out, the bees are more at ease. Only check on the hive when it is not raining, wind-blowing, or clouding.

- **Use a smoker:** A smoker prevents bees from automatically moving defensively by removing the defensive pheromones the bees tend to exude when they are in danger. Gentle puffs of smoke at the entrance and top of the hive should subdue the bees if they are aggressive.

- **Be Gentle when Handling the Bees:** You need to be sure of every single action you take. Do not do things dexterously by flinging your body, but try not to be too crazy and stiff-boned when dealing with the hive.

- **Wear the Necessary Attire:** Always put on your complete covering, including a veil, gloves, and the bee suit. This is not only for protection against stings but does give one assurance in dealing with offensive bees.

If aggression persists despite these measures, then returning the hive may be unavoidable. A cool-headed new queen will be able to tame the hot-headedness of the colony in due course.

Addressing and solving practical problems

In addition to swarming and aggressive behavior, there are other problems that beekeepers may face, and these include pests, queenlessness, and environmental stress. Not knowing how to deal with these challenges is bound to be injurious to the hive's survival.

Queenlessness

In the case of a hive with no queen, it is in a dire situation. The queen is the one who mainly fulfills the function of laying eggs, engaging in the activities that are required in the retaining structure of the colony. The death of the queen will lead to chaos in the hive, and the bees will work to rear another queen.

The symptoms of a colony not having a current queen include:

- Lack of eggs and larvae

- Increase in drone cells (the worker bees are beginning to deposit unfertilized eggs)

- Bees are getting annoyed or aggressive.

- If your suspicion that your hive is queenless is correct, you may bring in a new queen or let the bee's rear one if there are any queen cells. Often, however, being able to buy and insert a queen is the most effective and quickest option.

Pests and environmental stressors

Pests like Varroa mites, wax moths, and small hive beetles are known to destroy a bee colony.

Environmental factors such as pesticide use, inadequate feed, and poor per weather do affect the hive as well. To contend with these issues:

- **Regular Hive Inspections:** More frequent checks help to detect the pests early and act before much destruction is done.

- **Pest Management:** Use basic measures like offering some dust of powdered sugar, ala treats for varroa mites, or barrettes and traps for small hive beetles, among other methods of integrated pest management (IPM).

- **Nutritional Support:** Offer sufficient sources of forage for the bees, and if the natural food is not enough, include syrup made of sugars or pollen patties for them.

- **Shelter and protection:** Use insulation during the chilly season, protection from the sun on hot days, and ventilation in order to avoid condensation and other extreme environmental conditions that can harm the hive.

CHAPTER 11

BEYOND HONEY: EXPANDING YOUR BEEKEEPING HORIZONS

Even without any prior experience, many will discover that beekeeping goes beyond solely producing honey. Indeed, honey is the first incentive for most beekeepers; however, many other rather interesting and profitable activities can be taken once you grasp the basic rules. These could include increasing the size of your apiary by splitting hives and rearing the queens or making a successful enterprise out of the beekeeping and bee farming business.

Zealously indulging in the expansion of hives, rearing of queens, and the creation of a beekeeping enterprise that will make business sense will be discussed. We will also look into honey merchandising, product diversification strategies, and optimizing the use of bees. This is your guide to taking your beekeeping passion beyond honey and growing it into something even more fulfilling and financially rewarding.

Splitting hives and queen rearing

How to Expand Your Apiary

One of the most enjoyable and satisfying aspects of beekeeping is the growth of your apiary. Whether you are keeping bees as a hobby or are determined to go big for business after purchasing enough advances, splitting the hive is a procedure that helps in increasing the number of colonies in your care. It does not only assist in growing your business; it is also a wise method of minimizing swarming, which is a social habit of honeybee colonies that, when not properly controlled, may reduce the efficiency rate of the beehive.

Why Split Hives?

Increase Colony Numbers: Splitting your hive is a productive way of increasing your colony numbers since you do not have to spend money purchasing new bees. It's a cost-effective way to grow your apiary.

Prevent Swarming: Bees reproduce by division, in which the old or queen bee leaves the colony together with a few workers in search of a new colony. Sooner than later, when a colony becomes too big, the bees will eventually swarm in search of another colony. Splitting your hive helps prevent overcrowding and lowers the possibility of swarming.

Strengthen Weak Colonies: A split can help in reinforcing the weaker colonies by redistributing the biomass when brood frames and bees from a stronger colony are placed into the weaker one.

Speaking of brood nest division, all beginner beekeepers are certainly looking forward to hive splits at some point in the future.

- **Choose a Strong Hive:** The most suitable hive should be well populated with healthy bees and one that is rearing brood and making honey.

- **Prepare Equipment:** You are going to need some spare hive boxes, frames, and possibly the queen.

- **Move Frames:** Using the new hive box, place 3-5 brood and honey bearing frames along with the bees taken out of the old colony.

- **Introduce a Queen:** You may either have a queen bee introduced into the split or wait for the bees to make their queen, although this may take longer.

- **Monitor Both Hives:** Actively check up on the status of both the original and the split hives, ensuring that they are productive with eggs, bees, and other factors besides the population.

- Splitting hives not only helps you expand your apiary, but it's also an excellent way to learn more about bee behavior and colony dynamics. With experience, you'll become adept at identifying the right time to split a hive and managing the process successfully.

The Art of Raising Queen Bees

However, rearing queen bees entails some degree of both art and science, and it could uphold the presence of any beekeeper who might want to make an escalation in their apiary or enhance the genetic aspect of their colonies. To rear the queens, one is able to select and breed the queens that would better endow the colonies with desirable attributes such as high productivity, disease resistance, and good nature and temperament.

Why raise the Queen Bees?

- **Manage the Genetics of the Colony:** Building colonies of queens enables selective breeding of honeybees for desirable attributes such as being gentle or high honey-yielding.

- **Replace Worn-out Queens:** The normal life span of a queen is 2-3 years. By generating your own, replacement of older queens becomes easy, thus ensuring production in your hives.

- **Provide Other Beekeepers:** Once you have acquired the skill of raising the queens, there is an opportunity to sell good-quality queens to other farmers.

Steps to Rearing Queens

- **Select Larvae:** Larvae less than three days in age are taken from the strongest colonies of about the same age or older to preserve the desired qualities.

- **Create a Queenless Hive:** Isolate those larvae into queenless nuclei, so the colony of workers starts raising the larvae into queen bees.

- **Grafting:** A few of the beekeepers do use a grafting method where they place the larvae into specialized queen cups; however, some let the bees do it the natural way.

- **Nurture the Queen Cells:** After a queen cell has been capped, you may either put it in the hive where you want to keep the queen or you may use it to re-queen an existing colony.

- **Monitor Emergence:** After roughly 16 days following the introduction of new virgin queens, the new queens emerge, and their activity and fitness are evaluated.

Queen rearing is perhaps one of the most satisfying aspects of beekeeping, as it gives an opportunity to actively participate in the decisions about the future of the colonies. Though it takes a bit of learning and gaining experience, the benefits both for the beekeepers' business and for their wallets are really worth putting in some effort.

Turning beekeeping into a business

To those who are prepared to go further than the insect without commercialization phase, it is to be noted that the next phase would be taking up challenges in beekeeping in business scenarios. There are many opportunities to earn from beekeeping, including sales of honey, beeswax, and various hive products like propolis, royal jelly, and others. With proper organization and love for the insects, it would be possible to set up a profitable but sustainable beekeeping business.

Honey marketing and product diversification

It is widely accepted that honey is probably the most popular product associated with beekeeping. However, any other honey-based product and ancillary services require a keen bee farming business.

Honey Marketing

- **Know Your Market:** Knowing your market customers had to be the most critical step. Are the customers local markets or professional gourmet food stores? This is an advertising question and helps you focus your marketing activities.

- **Create a Brand:** In the contemporary market, with stiff competition for the consumer's attention, a strong brand is critical. Highlight your stress points on what makes you different, be it the organic honey, assorted creativity flavors from various flowers, or very attractive packaging

- **Utilize Online Platforms:** Many consumers nowadays also look for honey and other bee products on the internet. Set up a website that will help you sell your products, or do it using Etsy, Amazon, or local market delivery platforms.

- **Offer Tasting Events:** It is also highly recommended to organize tastings, especially during farmers' markets or in local shops where customers are likely to be interested in buying the honey and would want to know what type of honey is being sold.

Product diversification in a honey business includes the following:

- **Beeswax Products:** Very multifaceted and useful for candles as well as cosmetics and even natural polishes, a natural product known is beeswax. Making lip balm, lotions, and candles out of beeswax is another way of boosting your sales.

- **Propolis and Royal Jelly:** They are in products that provide nutrition to people. Propolis is normally present in the medicines that are directed internally with anti-inflammatory properties, and royal jelly is commonly used in dietary supplements. People who are concerned about their health will love it if you also provide these products along with your beekeeping business.

- **Pollination Services:** One more source of income is offering your bees to be hired out for pollination purposes. Especially farmers of fruits and vegetables tend to rent beehives from beekeepers in order to manage proper pollination of their crops.

How to Make the Most Out of Hive's Products: Maximize the Profits Derived Out of It

In order to gain the most profits, you need to be looking at all the possible sources of income from your hives. The greater the range, variety, and quality of the products and services, the more risk and uncertainty are taken out of the company's operations. Such areas include.

- **Sell it directly to customers themselves:** Several beekeepers have been successful in selling honey and honey and hive products directly by themselves at farmers' markets, craft shows, or through farmers' businesses.

- **Look at the wholesaling of honey:** Honey can be sold in bulk to some local retail chains, including restaurants, schools, and other businesses. Come over to the local shops that are likely to stock the honey and beeswax products you have.

- **Grow eCommerce:** Aside from the local market, target points include shipping honey as well as other bee products to people located in different parts of the country. With excellent packaging and the website being very friendly to the users, this can be a very profitable venture.

- **Conduct beekeeping lessons:** If you have the knowledge and skills of being a beekeeper, you can teach the interested people how to keep bees. This not only brings in revenue but also aids in societal building around the brand.

CONCLUSION

THE JOURNEY OF BEEKEEPING: FROM HOBBY TO PASSION

While initially forming and managing bees and hives can be considered a small-scale pastime, it is much more than that—it is a hobby with prospects of developing into a new adventure. Right from the time you make up your mind of putting up the first beehive, you are undertaking one of the oldest undertakings that man has involved himself/herself in. It is amazing how nurturing these insects in the colonies to flourish, harvest honey, pollinate blossoms, talk and work with the environment, and even dance is a very lovely and motivating task.

At the same time, your AC enters into activity all the major elements of beekeeping: insight into certain phases of life of a honey bee, managing hives during all the seasons, the honey extraction, and even some pest control measures. Mastering the craft of bee and each decisive stage of evolution, gradually taking steps towards more thorough and fine appreciation of the phenomenon of bee existence and the significance of bee activity in the biosphere.

In your second home, which is their beekeeping, it will metamorphose from a mere pastime to something more enthralling, an obsession. You will find yourself more engaged in the fascinating aspects of your bees' activities—knowing their cycles and taking actions that will determine your beekeeping setup in the future. What begins as a hobby in the backyard for a lot of hobbyists turns out to be a passion for a lifetime that is filled with experiences, learning, growth, and efforts in tending to nature.

While keeping bees, it is essential to cultivate patience, strong will, and assumptions of responsibility. Everything excellent and bad, all triumphs and disappointments, culminate in one thing: the pleasure of nurturing bees. Eventually, you will become addicted to them, the bees will care for their colonies, and the bees will enhance one's life.

It's quite astonishing how some new technologies are winning over the hearts of even the most stubborn tree professionals.

It can be self-gratifying to keep bees, but it can also point you towards a more compelling purpose the fight to save bees. As we have seen in this book, bees are general agricultural workers; as they help in the pollination of food, one out of three is contributed by these creatures. They are also critical for environmental conservation as they protect diversity, aid ecosystems, and buffer the provision of food resources across the globe.

Nevertheless, bees are losing their populations due to numerous challenges: loss of habitat, use of pesticides, changes in climate, and diseases. Because of these threats, the bee population started to dwindle, which could be devastating in every aspect and not only in the agricultural sector. Without bees, you will lack a fair amount of food, even damaging the balance of ecosystems.

As a keeper of bees, you are therefore at the forefront of assisting in the protection of these valuable insects. In as much as you operate the hives, feed the bees with the correct meals, and keep the place comfortable for them, all these work for their good. Each and every queen that you don, every colony that you manage in their brood nest is an endeavor in conserving bees for the future.

But your role doesn't stop with maintaining your own apiary. Beekeepers also are empowered to influence their communities. Explaining to people the role and impact of bees and the practices that protect these little creatures, as well as implementing such activities, will help raise awareness and foster engagement. Thus, if it is persuading the local community to use bee-friendly agricultural practices or even just showing the people in the surrounding area how to make pollinator pleasant areas such as gardens, your circle of influence can extend in such great ways.

Coming to embrace the Beekeeping Journey

As the novices embark on a beekeeping journey, let it be noted that this is a lifelong process, including much learning. No two seasons are ever the same, and there will always be a new problem waiting to be solved, be it pest management, swarm control, or changing weather conditions. But that is what is beautiful about beekeeping—it makes one active and inquisitive and, most importantly, ensures that he/she seeks knowledge for self-betterment.

Every year you do beekeeping will see you polonize, tilting you further and further into the deep end of knowledge. You will begin to pick the symptoms in a hive activity and be proficient in knowing the needs of the bees in that hive and knowing what signals they give. Observation is a mother of appreciation, and the more one appreciates their bees, the more beautiful and intelligent those bees are; hence, associations will be formed with them.

With time and as you keep getting better with beekeeping, you may gradually adopt a higher level of beekeeping. Maybe you'll choose to split different hives with the aim of increasing the number of colonies or even try to rear your own queens. You may also add value by venturing into other hive products like beeswax and bees royal jelly and making creative products like beeswax candles, lip balms, or healing salves.

For some people and their bees keeping hobbies and activities, it is not the end at one's backyard. These people go a step further and commercialize their interests by either harvesting honey and honey products for sale at farmers markets or offering pollination of local farms or instructional services on how to keep bees. This space is limitless, and there sure is room for new tilts and more exploration.

Concrete steps toward environmental responsibility

In keeping honey bees, we are not merely caregivers of the hives. We are active participants in the ecology. There is tremendous responsibility for creating or adopting altitudes toward certain particular hives and therefore influences detrimental to surrounding areas. Adopting sustainable approaches that welcome the presence of honey bees assists in the more significant vision of preserving the bees by ensuring that they do not become extinct.

This could imply that the use of organic or natural insect control methods, the planting of indigenous flowers that attract bees for nectar and pollen, or involvement in projects that build habitats for wild pollinators is an option. It could also mean banding together with other beekeepers and environmental groups to speak out for such policies as bans on harmful pesticides or support for pollinator-friendly agriculture that preserves pollinators on a bigger scale.

Everyone has a part to play in saving the bees: beekeepers, farmers, gardeners, and consumers. Together we can make thoughtful decisions that will guarantee that bees will still be with us, enabling us to grow our food, beautify the environment, and conserve biodiversity, which is the basis for all life on earth.

Look Ahead

On following this last orientation towards the conclusion of this paper, on turning this book and resuming your own work as a beekeeper, remember that every change, no matter how small, that arises as a result of any action is worth pursuing. Whether you are physically active in one hive or one thousand hives, you are doing a vital task for the bee colony's health and survival.

Beekeeping is also one such skill that grows with time. It is not just a mere activity of honey harvesting and taking care of the bees; it provides deeper engagement with nature, making one's input towards the environment much more purposeful and feeling oneself part of something bigger. It is about love; it is about cause; and it is about commitment.

And as you will be taking your next steps as a beekeeper, face all challenges, share all wins, and stay hungry for knowledge. The bees will be able to share with you more than you think you know, and in return, you will be offering them—and the planet a healthy, growing population bursting with life.

The transition from being a lighthearted hobbyist to a passionate beekeeper is a quest involving lots of trials, a lot of endurance and expenditure, and probably the most compelling benefits. May your beekeeping pursuits bring you a lot of satisfaction while, at the same time, helping us to save the incredible pollinators forever.

GLOSSARY

KEY TERMS USED IN THIS BOOK

Apiary

A location where beehives are kept, managed, and maintained. It's the official term for a bee farm or collection of hives.

Bee Bread

A mixture of pollen, honey, and bee secretions stored in the cells of the hive and used as food by bees.

Bee Space

A critical 3/8-inch gap within a hive that allows bees to move freely between the frames. If the gap is smaller, bees will seal it with propolis; if larger, they will fill it with comb.

Beeswax

A natural wax produced by worker bees from glands on their abdomen. It is used to build the honeycomb structure inside the hive.

Brood

Refers to the developing bees in the hive, including eggs, larvae, and pupae that are found in the brood cells of the hive.

Brood Box

The section of the hive where the queen lays her eggs and where the brood (developing bees) is kept.

Colony

The entire community of bees in a hive, including a queen, worker bees, and drones.

Comb (Honeycomb)

A structure made from beeswax where bees store honey, pollen, and brood. It consists of a series of hexagonal cells.

Drones

Male bees in a colony whose primary role is to mate with a queen. They do not collect nectar or pollen and do not have stingers.

Extractor

A machine used to remove honey from the comb by spinning the frames to force honey out of the cells.

Foulbrood

A bacterial disease that affects bee larvae and can cause significant damage to a colony. It is categorized into two types: American foulbrood (AFB) and European foulbrood (EFB).

Frames

Rectangular wooden structures that hold the comb inside the hive. They are removable, allowing beekeepers to inspect the hive and harvest honey.

Hive

A man-made structure used to house a bee colony. There are different types of hives, such as Langstroth, Warre, and Top-Bar.

Hive Inspection

The process of opening and examining the inside of a hive to monitor bee health, check for diseases, and ensure proper hive management.

Integrated Pest Management (IPM)

A strategy used to control pests in the hive through a combination of biological, chemical, and mechanical methods, minimizing harm to bees and the environment.

Langstroth Hive

A popular type of hive designed with removable frames, allowing for easy hive management and honey extraction.

Larva

The second stage in the life cycle of a bee. After hatching from the egg, the larva is fed and nurtured by worker bees until it pupates.

Nectar

A sweet liquid collected by bees from flowers, which they convert into honey through the process of evaporation and enzyme activity.

Nosema

A disease caused by a fungus that affects the digestive system of bees, weakening the colony.

Pollen

A fine powder collected from flowers by worker bees, which they mix with nectar to create bee bread. Pollen is a vital protein source for the colony.

Propolis

A sticky resin collected by bees from plants and used to seal gaps in the hive, providing insulation and protection against infections.

Queen Bee

The single reproductive female in a colony, responsible for laying all the eggs. The queen also emits pheromones that regulate the behavior and cohesion of the colony.

Royal Jelly

A protein-rich substance secreted by worker bees that is used to feed the queen and young larvae.

Smoker

A tool used by beekeepers to calm bees during hive inspections. It produces smoke, which masks alarm pheromones and reduces bee aggression.

Split (Splitting a Hive)

The process of dividing an existing colony into two or more separate colonies, often to increase the number of hives in an apiary.

Supers

Boxes added above the brood chamber in a hive where bees store surplus honey, which can be harvested by the beekeeper.

Swarm

A natural event in which a large group of bees, including the queen, leaves the hive to establish a new colony. Swarming occurs when the hive becomes overcrowded.

Swarm Prevention

Methods used by beekeepers to prevent swarming, such as providing additional space or splitting the colony.

Top-Bar Hive

A simple hive design that uses horizontal bars across the top of the hive for bees to build comb. It is often used by natural beekeepers due to its minimal intervention approach.

Varroa Mites

A parasitic mite that attaches to bees and weakens them by feeding on their bodily fluids. Varroa mites are one of the most significant threats to bee health worldwide.

Worker Bees

Female bees responsible for foraging, hive maintenance, brood care, and defending the colony. Workers are the majority of bees in a hive and perform all the essential tasks.

Warre Hive

A vertical top-bar hive that mimics the natural habitat of bees. It is often used by beekeepers who prefer a more hands-off, natural approach to hive management.

Wintering

The process of preparing a hive for winter, ensuring the bees have enough food and insulation to survive cold temperatures.